Combinatorics Through Guided Discovery

Combinatorics Through Guided Discovery

Kenneth P. Bogart
Dartmouth College

Editors

Mitchel T. Keller
Washington & Lee University

Oscar Levin
University of Northern Colorado

Kent E. Morrison
American Institute of Mathematics

Edition: First PreTeXt Edition

Website: http://bogart.openmathbooks.org/

© 2004 (main text); 2017 (Preface to PreTeXt Edition)   Kenneth P. Bogart (main text); Mitchel T. Keller, Oscar Levin, Kent E. Morrison (Preface to PreTeXt Edition)

Permission is granted to copy, distribute and/or modify this document under the terms of the GNU Free Documentation License, Version 1.3 or any later version published by the Free Software Foundation; with no Invariant Sections, no Front-Cover Texts, and no Back-Cover Texts. A copy of the license is included in the appendix entitled "GNU Free Documentation License".

About the Author

Kenneth Paul Bogart was born on October 6, 1943 in Cincinnati, OH. He graduated from Marietta College in Ohio in 1965, and earned his Ph.D. in mathematics from the California Institute of Technology in 1968. He married Ruth Tucker in 1966, and they moved to Hanover in 1968 where Ken was appointed an Assistant Professor of Mathematics. Ken remained in the job that he loved for 37 years being promoted to Associate Professor in 1974, and to Full Professor in 1980. Ken's career was characterized by a love of mathematics and scholarship, and a passion for teaching and mentoring at all levels within the mathematics curriculum. His passion for research is evidenced by over 60 journal articles and nine textbooks in his field of combinatorics. Ken's research covered a wide spectrum of topics within combinatorics.

Ken's mathematical roots were in algebra and lattice theory, and his earliest papers developed structural results for Noether lattices. One of the main topics in his research was partial orders, about which he wrote more than two dozen papers. This line of research started in the early 1970's with contributions to the theory of dimension for partial orders. A number of his papers treated applications of partial orders to the social sciences; for instance, he contributed to social choice theory by examining the optimal way to develop a consensus based on rankings that are partial orders. Interval orders and interval graphs played the most prominent role in Ken's research; his papers in this field span roughly thirty years, starting in the mid 1970's, and about half of his Ph.D. students worked in this area. Among his contributions in this area are the introduction and investigation of new concepts related to interval orders and graphs, the development of new and simpler proofs of key results, and the exploration of a number of structures that are natural variations or interesting special types of interval orders and graphs. Ken also contributed to the theory of error-correcting codes; in particular, he constructed a class of codes from partial orders. He collaborated on several papers in matroid theory, to which he contributed valuable insights from lattice theory and geometry.

Throughout the later part of his career, Ken became increasing interested in how students learn mathematics. His NSF-sponsored project of "guided-discovery" in combinatorics is an element that lives on in the math department. Ken also devoted a great deal of time to helping revise the teaching seminar which is fundamental part of the mathematics graduate program at Dartmouth.

For the past nine years, Ken and Ruth spent winters in Santa Rosa, CA, where they loved to hike and mountain bike.

This biography is excerpted from one written by T. R. Shemanske and posted on the website of the Department of Mathematics of Dartmouth College in April 2005. Shemanske wrote:

> I have borrowed freely from a number of published sources (below), and am especially grateful to Professor Joe Bonin for writing about Ken's research, Mary Pavone for her remarks about Ken's involvement with WISP, and for a photo from Ruth Bogart.
>
> Joe Bonin, The George Washington University; Mary Pavone, Director Women in Science Project; Press Democrat, Santa Rosa, CA (April 6, 2005); The Dartmouth, (April 4, 2005)

Preface

This book is an introduction to combinatorial mathematics, also known as combinatorics. The book focuses especially but not exclusively on the part of combinatorics that mathematicians refer to as "counting." The book consist almost entirely of problems. Some of the problems are designed to lead you to think about a concept, others are designed to help you figure out a concept and state a theorem about it, while still others ask you to prove the theorem. Other problems give you a chance to use a theorem you have proved. From time to time there is a discussion that pulls together some of the things you have learned or introduces a new idea for you to work with. Many of the problems are designed to build up your intuition for how combinatorial mathematics works. There are problems that some people will solve quickly, and there are problems that will take days of thought for everyone. Probably the best way to use this book is to work on a problem until you feel you are not making progress and then go on to the next one. Think about the problem you couldn't get as you do other things. The next chance you get, discuss the problem you are stymied on with other members of the class. Often you will all feel you've hit dead ends, but when you begin comparing notes and listening *carefully* to each other, you will see more than one approach to the problem and be able to make some progress. In fact, after comparing notes you may realize that there is more than one way to interpret the problem. In this case your first step should be to think together about what the problem is actually asking you to do. You may have learned in school that for every problem you are given, there is a method that has already been taught to you, and you are supposed to figure out which method applies and apply it. That is not the case here. Based on some simplified examples, you will discover the method for yourself. Later on, you may recognize a pattern that suggests you should try to use this method again.

The point of learning from this book is that you are learning how to discover ideas and methods for yourself, not that you are learning to apply methods that someone else has told you about. The problems in this book are designed to lead you to discover for yourself and prove for yourself the main ideas of combinatorial mathematics. There is considerable evidence that this leads to deeper learning and more understanding.

You will see that some of the problems are marked with bullets. Those are the problems that I feel are essential to having an understanding of what comes later, whether or not it is marked by a bullet. The problems with bullets are the problems in which the main ideas of the book are developed. Your instructor may leave out some of these problems because he or she plans not to cover future problems that rely on them. Many problems, in fact entire sections, are not marked in this way, because they use an important idea rather than developing one. Some other special symbols are described in what follows; a summary appears in the table below.

•	essential
○	motivational material
+	summary
⇒	especially interesting
*	difficult
·	essential for this section or the next

Some problems are marked with open circles. This indicates that they are designed to provide motivation for, or an introduction to, the important concepts, motivation with which some students may already be familiar. You will also see that some problems are marked with arrows. These point to problems that I think are particularly interesting. Some of them are also difficult, but not all are. A few problems that summarize ideas that have come before but aren't really essential are marked with a plus, and problems that are essential if you want to cover the section they are in or, perhaps, the next section, are marked with a dot (a small bullet). If a problem is relevant to a much later section in an essential way, I've marked it with a dot and a parenthetical note that explains where it will be essential. Finally, problems that seem unusually hard to me are marked with an asterisk. Some I've marked as hard only because I think they are difficult in light of what has come before, not because they are intrinsically difficult. In particular, some of the problems marked as hard will not seem so hard if you come back to them after you have finished more of the problems.

If you are taking a course, your instructor will choose problems for you to work on based on the prerequisites for and goals of the course. If you are reading the book on your own, I recommend that you try all the problems in a section you want to cover. Try to do the problems with bullets, but by all means don't restrict yourself to them. Often a bulleted problem makes more sense if you have done some of the easier motivational problems that come before it. If, after you've tried it, you want to skip over a problem without a bullet or circle, you should not miss out on much by not doing that problem. Also, if you don't find the problems in a section with no bullets interesting, you can skip them, understanding that you may be skipping an entire branch of combinatorial mathematics! And no matter what, read the textual material that comes before, between, and immediately after problems you are working on!

One of the downsides of how we learn math in high school is that many of us come to believe that if we can't solve a problem in ten or twenty minutes, then we can't solve it at all. There will be problems in this book that take hours of hard thought. Many of these problems were first conceived and solved by professional mathematicians, and *they* spent days or weeks on them. How can you be expected to solve them at all then? You have a context in which to work, and even though some of the problems are so open ended that you go into them without any idea of the answer, the context and the leading examples that preceded them give you a structure to work with. That doesn't mean you'll get them right away, but you will find a real sense of satisfaction when you see what you can figure out with concentrated thought. Besides, you can get hints!

Some of the questions will appear to be trick questions, especially when you get the answer. They are not intended as trick questions at all. Instead they are designed so that they don't tell you the answer in advance. For example the answer to a question that begins "How many..." might be "none." Or there might be just one example (or even no examples) for a problem that asks you to find all examples of something. So when you read a question, unless it directly tells you what the answer is and asks you to show it is true, don't expect the wording of the problem to suggest the answer. The book isn't designed this way to be cruel. Rather, there is evidence that the more open-ended a question is, the more deeply you learn from working on it. If you do go on to do mathematics later in life, the problems that come to you from the real world or from exploring a mathematical topic are going to be open-ended problems because nobody will have done them before. Thus working on open-ended problems now should help to prepare you to do mathematics later on.

You should try to write up answers to all the problems that you work on. If you claim something is true, you should explain why it is true; that is you should prove it. In some cases an idea is introduced before you have the tools to prove it, or the proof of something will add nothing to your understanding. In such problems there is a remark telling you not to bother with a proof. When you write up a problem, remember that the instructor has to be able to "get" your ideas and understand exactly what you are saying. Your instructor is going to choose some of your solutions to read carefully and give you detailed feedback on. When you get this feedback, you should think it over carefully and then write the solution again! You may be asked not to have someone else read your solutions to some of these problems until your instructor has. This is so that the instructor can offer help which is aimed at your needs. On other problems it is a good idea to seek feedback from other students. One of the best ways of learning to write clearly is to have someone point out to you where it is hard to figure out what you mean. The crucial thing is to make it clear to your reader that you really want to know where you may have left something out, made an unclear statement, or failed to support a statement with a proof. It is often very helpful to choose people who have not yet become an expert with the problems, as long as they realize it will help you most for them to tell you about places in your solutions they do not understand, even if they think it is their problem and not yours!

As you work on a problem, think about why you are doing what you are doing. Is it helping you? If your current approach doesn't feel right, try to see why. Is this a problem you can decompose into simpler problems? Can you see a way to make up a simple example, even a silly one, of what the problem is asking you to do? If a problem is asking you to do something for every value of an integer n, then what happens with simple values of n like 0, 1, and 2? Don't worry about making mistakes; it is often finding mistakes that leads mathematicians to their best insights. Above all, don't worry if you can't do a problem. Some problems are given as soon as there is one technique you've learned that might help do that problem. Later on there may be other techniques that you can bring back to that problem to try again. The notes have been designed this way on purpose. If you happen to get a hard problem with the bare minimum of tools, you will have accomplished much. As you go along, you will see your ideas appearing again later in other problems. On the other hand, if you don't get the problem the first time through, it will be nagging at you as you work on other things, and when you see the idea for an old problem in new work, you will know you are learning.

There are quite a few concepts that are developed in this book. Since most of the intellectual content is in the problems, it is natural that definitions of concepts will often be within problems. When you come across an unfamiliar term in a problem, it is likely it was defined earlier. Look it up in the index, and with luck (hopefully no luck will really be needed!) you will be able to find the definition.

Above all, this book is dedicated to the principle that doing mathematics is fun. As long as you know that some of the problems are going to require more than one attempt before you hit on the main idea, you can relax and enjoy your successes, knowing that as you work more and more problems and share more and more ideas, problems that seemed intractable at first become a source of satisfaction later on.

The development of this book is supported by the National Science Foundation. An essential part of this support is an advisory board of faculty members from a wide variety of institutions who have made valuable contributions. They are Karen Collins, Wesleyan University, Marc Lipman, Indiana University/Purdue University, Fort Wayne, Elizabeth MacMahon, Lafayette College, Fred McMorris, Illinois Institute of Technology, Mark Miller, Marietta College, Rosa Orellana, Dartmouth College, Vic Reiner, University of Minnesota, and Lou Shapiro, Howard University. The overall design and most of the problems in the appendix on exponential generating functions are due to Professors Reiner and Shapiro. Any errors or confusing writing in that appendix are due to me! I believe the board has managed both to make the book more accessible and more interesting.

Preface to PreTeXt edition

At the time of his death in 2005, Ken Bogart was working on this NSF-supported effort to create a combinatorics textbook that developed the key ideas of undergraduate combinatorics through "guided discovery", or what many today typically call inquiry-based learning. The project was under contract with Springer-Verlag for a commercially-published print edition, but Ken's untimely passing left the project in an unfinished state. Bogart's family asked the Department of Mathematics at Dartmouth College, where he had spent his entire career after earning his Ph.D. from Caltech in 1968, to distribute the text freely under the GNU Free Documentation License. This open-source release came with some notes, however. Those notes, listed on the book's Dartmouth page, were:

1. The contents of the archive are released under the terms of the GNU Free Documentation License (FDL), a copy of which is contained in the archive.
2. The contents of the archive are released in "as is" condition, which in particular means that the state of the source files is not in agreement with the pdf versions of the text. A README offers some guidance.
3. Many people have already used the textbook in courses at various universities throughout the country. It is the hope of the Bogart family that this project continues to grow to completion with the efforts of those who download this archive.

The caveat in the second note seemed to be the largest toward fulfilling the goal of the third, as the "official" version of the PDF had a different chapter structure than the LaTeX source files provided. This was mostly the result of splitting a chapter into two and rearranging a few topics, but there were also places where problems were split or merged between the source version and the PDF version. The PDF version also came with copious hints that readers could access online, but no LaTeX source existed for these hints.

This PreTeXt edition of *Combinatorics through Guided Discovery* attempts to help fulfill the Bogart family's wish to see the project grow and reach a complete state. One of us (MTK) had used the official PDF to teach a combinatorics course in Winter 2015 and mentioned this fact at a workshop on open source textbooks and PreTeXt (then MathBook XML) organized by the American Institute of Mathematics in the spring of 2016. This caught the attention of KEM, since *Combinatorics through Guided*

Discovery had been placed on the AIM list of approved open source textbooks, but there had been no success in finding someone to take on the task of updating the source to match the PDF. The three of us came together again in May 2017 at the University of Puget Sound for another workshop on open educational resources and agreed to cooperate to complete the conversion of this book to PreTeXt. The fact that OL wanted to use parts of the book for his Fall 2017 class gave us the motivation required to complete the project over the summer.

For this edition, our goal has been to reproduce the text of Bogart's final PDF as faithfullly as possible. Based on our own classroom uses, we have notes about problems that could use revising, but we agreed the right first step would be to have source files that matched what Bogart left. We have, however, corrected obvious errors along the way, which included moving the Supplementary Chapter Problems in Chapter 3 from the level of a subsection to the level of a section for consistency with the other chapters. Footnotes may be numbered differently, as in this edition, a footnote in the body of a problem is rendered with the problem and numbered in a different sequence. The hints that previously were accessed by links from the PDF to a Dartmouth webpage have also been included in the backmatter of the print edition as Appendix D. Links to open hints in place are available in the HTML version, while in the print and PDF edition, the existence of a hint is indicated by "(h)" at the end of the problem (or part of a problem). David Farmer provided invaluable assistance by automating the initial conversion of LaTeX files to PreTeXt and extracting the text of the hints from the PDF files. We then worked in parallel to compare the official PDF to what we were able to produce from the source until they matched. Since this process could not be truly automated, we suspect there will be some places where Bogart's PDF and this edition do not match. We welcome reports of these through issues and pull requests on the Github site for the book https://github.com/OpenDiscreteMath/ibl-combinatorics/. Going forward, we would like to see community-driven updates to further develop the text, either by improving existing problems, adding new problems on existing topics, or adding new topics suitable for a course based on this text. One area of development may be to include SageMath to the text, since PreTeXt includludes a number of nice features for doing this and some of the material may benefit from the addition of a computer algebra system to allow more interesting calculations than would be feasible by hand.

An HTML version of this text is available at http://bogart.openmathbooks.org/. A low-cost print edition is available for purchase online. The cost of the print edition is kept as low as possible, and any royalties received support costs associated with hosting and distributing the text. A PDF copy of print edition is also posted on the book's site. The PDF may provide a better experience for searching than the HTML version.

Mitchel T. Keller, Oscar Levin, and Kent E. Morrison
Lexington, Virginia; Greeley, Colorado; and San Jose, California
December 2017

Contents

About the Author v

Preface vii

Preface to PreTeXt edition xi

1 **What is Combinatorics?** 1
   1.1 About These Notes 1
   1.2 Basic Counting Principles 2
      1.2.1 The sum and product principles 5
      1.2.2 Functions and directed graphs 9
      1.2.3 The bijection principle 11
      1.2.4 Counting subsets of a set 11
      1.2.5 Pascal's Triangle 12
      1.2.6 The quotient principle 15
   1.3 Some Applications of the Basic Principles 20
      1.3.1 Lattice paths and Catalan Numbers 20
      1.3.2 The Binomial Theorem 23
      1.3.3 The pigeonhole principle 25
      1.3.4 Ramsey Numbers 26
   1.4 Supplementary Chapter Problems 28

2 **Applications of Induction and Recursion in Combinatorics and Graph Theory** 31
   2.1 Some Examples of Mathematical Induction 31
      2.1.1 Mathematical induction 31
      2.1.2 Binomial Coefficients and the Binomial Theorem 33
      2.1.3 Inductive definition 33
      2.1.4 Proving the general product principle (Optional) 34
      2.1.5 Double Induction and Ramsey Numbers 35
      2.1.6 A bit of asymptotic combinatorics 36
   2.2 Recurrence Relations 37
      2.2.1 Examples of recurrence relations 38
      2.2.2 Arithmetic Series (optional) 39
      2.2.3 First order linear recurrences 40
      2.2.4 Geometric Series 40
   2.3 Graphs and Trees 41

	2.3.1	Undirected graphs	41
	2.3.2	Walks and paths in graphs	43
	2.3.3	Counting vertices, edges, and paths in trees	43
	2.3.4	Spanning trees	45
	2.3.5	Minimum cost spanning trees	46
	2.3.6	The deletion/contraction recurrence for spanning trees	47
	2.3.7	Shortest paths in graphs	48
2.4	Supplementary Problems	49

3 Distribution Problems — 51

3.1	The idea of a distribution	51
	3.1.1	The twentyfold way	51
	3.1.2	Ordered functions	54
	3.1.3	Multisets	56
	3.1.4	Compositions of integers	56
	3.1.5	Broken permutations and Lah numbers	57
3.2	Partitions and Stirling Numbers	58
	3.2.1	Stirling Numbers of the second kind	58
	3.2.2	Stirling Numbers and onto functions	60
	3.2.3	Stirling Numbers and bases for polynomials	61
3.3	Partitions of Integers	63
	3.3.1	The number of partitions of k into n parts	63
	3.3.2	Representations of partitions	63
	3.3.3	Ferrers and Young Diagrams and the conjugate of a partition	64
	3.3.4	Partitions into distinct parts	69
3.4	Supplementary Problems	70

4 Generating Functions — 73

4.1	The Idea of Generating Functions	73
	4.1.1	Visualizing Counting with Pictures	73
	4.1.2	Picture functions	74
	4.1.3	Generating functions	75
	4.1.4	Power series	77
	4.1.5	Product principle for generating functions	78
	4.1.6	The extended binomial theorem and multisets	79
4.2	Generating functions for integer partitions	81
4.3	Generating Functions and Recurrence Relations	85
	4.3.1	How generating functions are relevant	85
	4.3.2	Fibonacci numbers	86
	4.3.3	Second order linear recurrence relations	86
	4.3.4	Partial fractions	87
	4.3.5	Catalan Numbers	89
4.4	Supplementary Problems	90

5 The Principle of Inclusion and Exclusion — 93
- 5.1 The size of a union of sets 93
  - 5.1.1 Unions of two or three sets 93
  - 5.1.2 Unions of an arbitrary number of sets 94
  - 5.1.3 The Principle of Inclusion and Exclusion 95
- 5.2 Application of Inclusion and Exclusion 97
  - 5.2.1 Multisets with restricted numbers of elements 97
  - 5.2.2 The Menage Problem 97
  - 5.2.3 Counting onto functions 97
  - 5.2.4 The chromatic polynomial of a graph 98
- 5.3 Deletion-Contraction and the Chromatic Polynomial 99
- 5.4 Supplementary Problems 101

6 Groups acting on sets — 103
- 6.1 Permutation Groups 103
  - 6.1.1 The rotations of a square 103
  - 6.1.2 Groups of Permutations 104
  - 6.1.3 The symmetric group 106
  - 6.1.4 The dihedral group 107
  - 6.1.5 Group tables (Optional) 110
  - 6.1.6 Subgroups 111
  - 6.1.7 The cycle structure of a permutation 112
- 6.2 Groups Acting on Sets 114
  - 6.2.1 Groups acting on colorings of sets 116
  - 6.2.2 Orbits 119
  - 6.2.3 The Cauchy-Frobenius-Burnside Theorem 122
- 6.3 Pólya-Redfield Enumeration Theory 125
  - 6.3.1 The Orbit-Fixed Point Theorem 126
  - 6.3.2 The Pólya-Redfield Theorem 128
- 6.4 Supplementary Problems 131

A Relations — 133
- A.1 Relations as sets of Ordered Pairs 133
  - A.1.1 The relation of a function 133
  - A.1.2 Directed graphs 135
  - A.1.3 Digraphs of Functions 136
- A.2 Equivalence relations 138

B Mathematical Induction — 145
- B.1 The Principle of Mathematical Induction 145
  - B.1.1 The ideas behind mathematical induction 145
  - B.1.2 Mathematical induction 147
  - B.1.3 Proving algebraic statements by induction 148
  - B.1.4 Strong Induction 149

C	**Exponential Generating Functions**	**151**
	C.1 Indicator Functions	151
	C.2 Exponential Generating Functions	152
	C.3 Applications to recurrences.	154
	C.3.1 Using calculus with exponential generating functions	155
	C.4 The Product Principle for EGFs	156
	C.5 The Exponential Formula	162
	C.6 Supplementary Problems	166
D	**Hints to Selected Problems**	**169**
E	**GNU Free Documentation License**	**193**
	Index	**199**

Chapter 1

What is Combinatorics?

Combinatorial mathematics arises from studying how we can *combine* objects into arrangements. For example, we might be combining sports teams into a tournament, samples of tires into plans to mount them on cars for testing, students into classes to compare approaches to teaching a subject, or members of a tennis club into pairs to play tennis. There are many questions one can ask about such arrangements of objects. Here we will focus on questions about *how many ways* we may combine the objects into arrangements of the desired type. These are called **counting problems**. Sometimes, though, combinatorial mathematicians ask if an arrangement is possible (if we have ten baseball teams, and each team has to play each other team once, can we schedule all the games if we only have the fields available at enough times for forty games?). Sometimes they ask if all the arrangements we might be able to make have a certain desirable property (Do all ways of testing 5 brands of tires on 5 different cars [with certain additional properties] compare each brand with each other brand on at least one common car?). Problems of these sorts come up throughout physics, biology, computer science, statistics, and many other subjects. However, to demonstrate all these relationships, we would have to take detours into all these subjects. While we will give some important applications, we will usually phrase our discussions around everyday experience and mathematical experience so that the student does not have to learn a new context before learning mathematics in context!

1.1 About These Notes

These notes are based on the philosophy that you learn the most about a subject when you are figuring it out directly for yourself, and learn the least when you are trying to figure out what someone else is saying about it. On the other hand, there is a subject called combinatorial mathematics, and that is what we are going to be studying, so we will have to tell you some basic facts. What we are going to try to do is to give you a chance to discover many of the interesting examples that usually appear as textbook examples and discover the principles that appear as textbook theorems. Your main activity will be solving problems designed to lead you to discover the basic principles of combinatorial mathematics. Some of the problems lead you through a new idea, some give you a chance to describe what you have learned in a sequence of problems, and some are quite challenging. When you find a problem challenging, don't give up on it, but don't let it stop you from going on with other problems. Frequently you will find an idea in a later problem that you can take back to the one you skipped over or only partly finished in order to finish it off. With that in mind, let's get started. In the problems that follow, you will see some problems marked on the left with various symbols. The preface gives a full explanation of these symbols and discusses in greater detail why the book is organized as it is! Table 1.1.1, which is repeated from the preface, summarizes the meaning of the symbols.

•	essential
○	motivational material
+	summary
⇒	especially interesting
*	difficult
·	essential for this section or the next

Table 1.1.1: The meaning of the symbols to the left of problem numbers.

1.2 Basic Counting Principles

○ **Problem 1.** Five schools are going to send their baseball teams to a tournament, in which each team must play each other team exactly once. How many games are required? (h) ~~15~~ 10

• **Problem 2.** Now some number n of schools are going to send their baseball teams to a tournament, and each team must play each other team exactly once. Let us think of the teams as numbered 1 through n.

  $n-1$ (a) How many games does team 1 have to play in?

  $n-2$ (b) How many games, other than the one with team 1, does team two have to play in?

  $n-i$ (c) How many games, other than those with the first $i-1$ teams, does team i have to play in?

  ? (d) In terms of your answers to the previous parts of this problem, what is the total number of games that must be played?

• **Problem 3.** One of the schools sending its team to the tournament has to send its players from some distance, and so it is making sandwiches for team members to eat along the way. There are three choices for the kind of bread and five choices for the kind of filling. How many different kinds of sandwiches are available? (h)

15

Problem 4. An **ordered pair** (a, b) consists of two things we call a and b. We say a is the first member of the pair and b is the second member of the pair. If M is an m element set and N is an n-element set, how many ordered pairs are there whose first member is in M and whose second member is in N? Does this problem have anything to do with any of the previous problems?

MN

Problem 5. Since a sandwich by itself is pretty boring, students from the school in Problem 3 are offered a choice of a drink (from among five different kinds), a sandwich, and a fruit (from among four different kinds). In how many ways may a student make a choice of the three items now?

15 × 5 × 4 = 300

Problem 6. The coach of the team in Problem 3 knows of an ice cream parlor along the way where she plans to stop to buy each team member a triple decker cone. There are 12 different flavors of ice cream, and triple decker cones are made in homemade waffle cones. Having chocolate ice cream as the bottom scoop is different from having chocolate ice cream as the top scoop. How many possible ice cream cones are going to be available to the team members? How many cones with three different kinds of ice cream will be available? (h)

$12^3 = 1728$

$12 × 11 × 10 = 1320$

Problem 7. The idea of a function is ubiquitous in mathematics. A function f from a set S to a set T is a relationship between the two sets that associates exactly one member $f(x)$ of T with each element x in S. We will come back to the ideas of functions and relationships in more detail and from different points of view from time to time. However, the quick review above should probably let you answer these questions. If you have difficulty with them, it would be a good idea to go now to Appendix A and work through Section A.1 which covers this definition in more detail. You might also want to study Section A.1.3 to learn to visualize the properties of functions. We will take up the topic of this section later in this chapter as well, but in less detail than is in the appendix.

  (a) Using $f, g, \ldots,$ to stand for the various functions, write down all the different functions you can from the set $\{1,2\}$ to the set $\{a,b\}$.

1. What is Combinatorics?

$f(a) = 1 \quad f(a) = \frac{1}{3}$
$f(b) = 2 \quad f(b) = \frac{1}{3} \quad = \frac{1}{9}$

2. For example, you might start with $f(1) = a$, $f(2) = b$. How many functions are there from the set $\{1, 2\}$ to the set $\{a, b\}$? (h)

$f(1) = a \quad f(1) = a$
$f(2) = a \quad f(2) = b$
$f(3) = b \quad f(3) = b$

$f(1) = b \quad f(1) = b$
$f(2) = b \quad f(2) = a$
$f(3) = a \quad f(3) = a$

$f(1) = b \quad \begin{matrix} a \\ b \\ a \end{matrix}$
$\quad\quad a$
$\quad\quad b$

(b) How many functions are there from the three element set $\{1, 2, 3\}$ to the two element set $\{a, b\}$? (h) $\cancel{4} \quad \cancel{6} \quad \frac{1}{2} \times \frac{1}{2} \times \frac{1}{2} = 8$

(c) How many functions are there from the two element set $\{a, b\}$ to the three element set $\{1, 2, 3\}$? (h) 9

(d) How many functions are there from a three element set to a 12 element set? $\cancel{144} \quad \frac{1}{12} \times \frac{1}{12^3} = 12^3$

(e) The function f is called **one-to-one** or an **injection** if whenever x is different from y, $f(x)$ is different from $f(y)$. How many one-to-one functions are there from a three element set to a 12 element set?

(f) Explain the relationship between this problem and Problem 6.

Set $S = \{1, 2, 3\}$ Set $T = \{1 \ldots 12\}$
3 inputs   12 possible outputs

- **Problem 8.** A group of hungry team members in Problem 6 notices it would be cheaper to buy three pints of ice cream for them to split than to buy a triple decker cone for each of them, and that way they would get more ice cream. They ask their coach if they can buy three pints of ice cream.

  (a) In how many ways can they choose three pints of different flavors out of the 12 flavors? (h)

  (b) In how many ways may they choose three pints if the flavors don't have to be different? (h)

$\sum e_{\text{sets}}$

- **Problem 9.** Two sets are said to be **disjoint** if they have no elements in common. For example, $\{1, 3, 12\}$ and $\{6, 4, 8, 2\}$ are disjoint, but $\{1, 3, 12\}$ and $\{3, 5, 7\}$ are not. Three or more sets are said to be **mutually disjoint** if no two of them have any elements in common. What can you say about the size of the union of a finite number of finite (mutually) disjoint sets? Does this have anything to do with any of the previous problems?

- **Problem 10.** Disjoint subsets are defined in Problem 9. What can you say about the size of the union of m (mutually) disjoint sets, each of size n? Does this have anything to do with any of the previous problems?

$n \times m$

1.2.1 The sum and product principles

These problems contain among them the kernels of many of the fundamental ideas of combinatorics. For example, with luck, you just stated the sum principle (illustrated in Figure 1.2.1), and product principle (illustrated in Figure 1.2.2) in Problems 9 and Problem 10. These two counting principles are the basis on which we will develop many other counting principles.

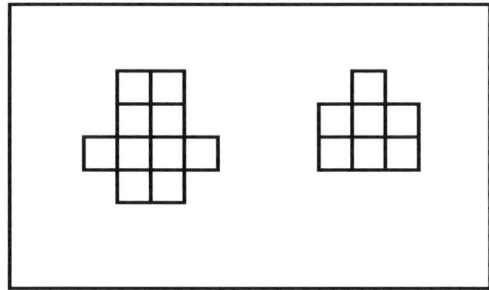

Figure 1.2.1: The union of these two disjoint sets has size 17.

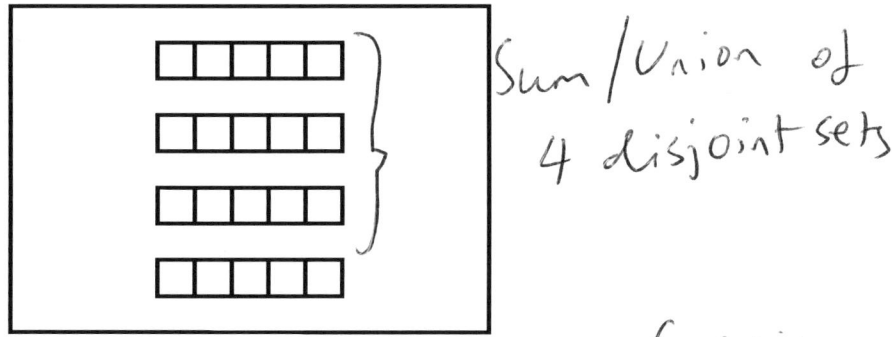

Figure 1.2.2: The union of four disjoint sets of size five.

You may have noticed some standard mathematical words and phrases such as **set**, **ordered pair**, **function** and so on creeping into the problems. One of our goals in these notes is to show how most counting problems can be recognized as counting all or some of the elements of a set of standard mathematical objects. For example Problem 4 is meant to suggest that the question we asked in Problem 3 was really a problem of counting all the ordered pairs consisting of a bread choice and a filling choice. We use $A \times B$ to stand for the set of all ordered pairs whose first element is in A and whose second element is in B and we call $A \times B$ the **Cartesian product** of A and B, so you can think of Problem 4 as asking you for the size of the Cartesian product of M and N, that is, asking you to count the number of elements of this Cartesian product.

When a set S is a union of disjoint sets $B_1, B_2, \ldots, B_m$ we say that the sets $B_1, B_2, \ldots, B_m$ are a **partition** of the set S. Thus a partition of S is a (special kind of) set of sets. So that we don't find ourselves getting confused between the set S

and the sets B_i into which we have divided it, we often call the sets $B_1, B_2, \ldots, B_m$ the **blocks** of the partition. In this language, the **sum principle** says that

 if we have a partition of a set S, then the size of S is the sum of the sizes of the blocks of the partition.

The **product principle** says that

 if we have a partition of a set S into m blocks, each of size n, then S has size mn.

You'll notice that in our formal statement of the sum and product pinciple we talked about a partition of a finite set. We could modify our language a bit to cover infinite sizes, but whenever we talk about sizes of sets in what follows, we will be working with finite sets. So as to avoid possible complications in the future, let us agree that when we refer to the size of a set, we are implicitly assuming the set is finite. There is another version of the product principle that applies directly in problems like Problem 5 and Problem 6, where we were not just taking a union of m disjoint sets of size n, but rather m disjoint sets of size n, each of which was a union of m' disjoint sets of size n'. This is an inconvenient way to have to think about a counting problem, so we may rephrase the product principle in terms of a sequence of decisions:

- **Problem 11.** If we make a sequence of m choices for which

    - there are k_1 possible first choices, and

    - for each way of making the first $i - 1$ choices, there are k_i ways to make the ith choice,

    then in how many ways may we make our sequence of choices? (You need not prove your answer correct at this time.)

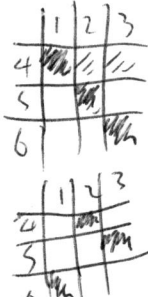

The counting principle you gave in Problem 11 is called the **general product principle**. We will outline a proof of the general product pinciple from the original product principle in Problem 80. Until then, let us simply accept it as another counting principle. For now, notice how much easier it makes it to explain why we multiplied the things we did in Problem 5 and Problem 6.

⇒ **Problem 12.** A tennis club has $2n$ members. We want to pair up the members by twos for singles matches.

(a) In how many ways may we pair up all the members of the club? (Hint: consider the cases of 2, 4, and 6 members.) (h)

(b) Suppose that in addition to specifying who plays whom, for each pairing we say who serves first. Now in how many ways may we specify our pairs? $2n^2$

+ **Problem 13.** Let us now return to Problem 7 and justify—or perhaps finish—our answer to the question about the number of functions from a three-element set to a 12-element set.

   (a) How can you justify your answer in Problem 7 to the question "How many functions are there from a three element set (say $[3] = \{1,2,3\}$) to a twelve element set (say $[12]$)?"

   (b) Based on the examples you've seen so far, make a conjecture about how many functions there are from the set
   $$[m] = \{1, 2, 3, \ldots, m\}$$
   to $[n] = \{1, 2, 3, \ldots, n\}$ and prove it.

   (c) A common notation for the set of all functions from a set M to a set N is N^M. Why is this a good notation? succint

+ **Problem 14.** Now suppose we are thinking about a set S of functions f from $[m]$ to some set X. (For example, in Problem 6 we were thinking of the set of functions from the three possible places for scoops in an ice-cream cone to 12 flavors of ice cream.) Suppose there are k_1 choices for $f(1)$. (In Problem 6, k_1 was 12, because there were 12 ways to choose the first scoop.) Suppose that for each choice of $f(1)$ there are k_2 choices for $f(2)$. (For example, in Problem 6 k_2 was 12 if the second flavor could be the same as the first, but k_2 was 11 if the flavors had to be different.) In general, suppose that for each choice of $f(1), f(2), \ldots f(i-1)$, there are k_i choices for $f(i)$. (For example, in Problem 6, if the flavors have to be different, then for each choice of $f(1)$ and $f(2)$, there are 10 choices for $f(3)$.)
What we have assumed so far about the functions in S may be summarized as

   - There are k_1 choices for $f(1)$.
   - For each choice of $f(1), f(2), \ldots, f(i-1)$, there are k_i choices for $f(i)$.

   How many functions are in the set S? Is there any practical difference between the result of this problem and the general product principle?

The point of Problem 14 is that the general product principle can be stated informally, as we did originally, or as a statement about counting sets of standard concrete mathematical objects, namely functions.

⇒ **Problem 15.** A roller coaster car has n rows of seats, each of which has room for two people. If n men and n women get into the car with a man and a woman in each row, in how many ways may they choose their seats? (h)

$\{3\} \subseteq \{n\}$
N^M

- **Problem 16.** How does the general product principle apply to Problem 6?

k^n

- **Problem 17.** In how many ways can we pass out k distinct pieces of fruit to n children (with no restriction on how many pieces of fruit a child may get)?

- **Problem 18.** How many subsets does a set S with n elements have? (h)

o **Problem 19.** Assuming $k \leq n$, in how many ways can we pass out k distinct pieces of fruit to n children if each child may get at most one? What is the number if $k > n$? Assume for both questions that we pass out all the fruit. (h)

- **Problem 20.** Another name for a list, in a specific order, of k distinct things chosen from a set S is a k-**element permutation** of S. We can also think of a k-element permutation of S as a one-to-one function (or, in other words, injection) from $[k] = \{1, 2, \ldots, k\}$ to S. How many k-element permutations does an n-element set have? (For this problem it is natural to assume $k \leq n$. However the question makes sense even if $k > n$. What is the number of k-element permutations of an n-element set if $k > n$? (h)

There are a number of different notations for the number of k-element permutations of an n-element set. The one we shall use was introduced by Don Knuth; namely $n^{\underline{k}}$, read "n to the k falling" or "n to the k down". In Problem 20 you may have shown that

$$n^{\underline{k}} = n(n-1)\cdots(n-k+1) = \prod_{i=1}^{k}(n-i+1). \quad (1.1)$$

It is standard to call $n^{\underline{k}}$ the k-**th falling factorial power of** n, which explains why we use exponential notation. Of course we call it a **factorial** power since $n^{\underline{n}} = n(n-1)\cdots 1$ which we call n-**factorial** and denote by $n!$. If you are unfamiliar with the Π notation, or **product notation** we introduced for products in Equation (1.1), it works just like the Σ notation works for summations.

$\dfrac{n!}{(n-k)!}$

- **Problem 21.** Express $n^{\underline{k}}$ as a quotient of factorials.

$\Rightarrow$ **Problem 22.** How should we define $n^{\underline{0}}$? 1

1.2.2 Functions and directed graphs

As another example how standard mathematical language relates to counting problems, Problem 7 explicitly asked you to relate the idea of counting functions to the question of Problem 6. You have probably learned in algebra or calculus how to draw graphs in the Cartesian plane of functions from a set of numbers to a set of numbers. You may recall how we can determine whether a graph is a graph of a function by examining whether each vertical straight line crosses the graph at most one time. You might also recall how we can determine whether such a function is one-to-one by examining whether each horizontal straight line crosses the graph at most one time. The functions we deal with will often involve objects which are not numbers, and will often be functions from one finite set to another. Thus graphs in the cartesian plane will often not be available to us for visualizing functions.

However, there is another kind of graph called a **directed graph** or **digraph** that is especially useful when dealing with functions between finite sets. We take up this topic in more detail in Appendix A, particularly Subsection A.1.2 and Subsection A.1.3. In Figure 1.2.3 we show several examples of digraphs of functions.

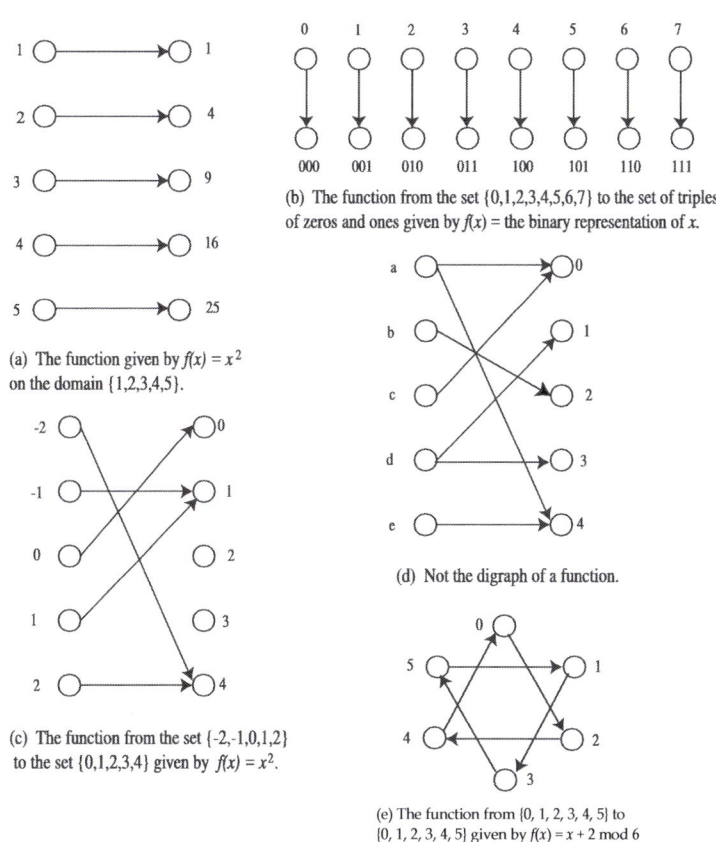

(a) The function given by $f(x) = x^2$ on the domain $\{1,2,3,4,5\}$.

(b) The function from the set $\{0,1,2,3,4,5,6,7\}$ to the set of triples of zeros and ones given by $f(x)$ = the binary representation of x.

(c) The function from the set $\{-2,-1,0,1,2\}$ to the set $\{0,1,2,3,4\}$ given by $f(x) = x^2$.

(d) Not the digraph of a function.

(e) The function from $\{0, 1, 2, 3, 4, 5\}$ to $\{0, 1, 2, 3, 4, 5\}$ given by $f(x) = x + 2 \bmod 6$

Figure 1.2.3: What is a digraph of a function?

A directed graph for visualising functions

If we have a function f from a set S to a set T, we draw a line of dots or circles, called **vertices** to represent the elements of S and another (usually parallel) line of circles or dots to represent the elements of T. We then draw an arrow from the circle for x to the circle for y if $f(x) = y$. Sometimes, as in part (e) of the figure, if we have a function from a set S to itself, we draw only one set of vertices representing the elements of S, in which case we can have arrows both entering and leaving a given vertex. As you see, the digraph can be more enlightening in this case if we experiment with the function to find a nice placement of the vertices rather than putting them in a row.

Notice that there is a simple test for whether a digraph whose vertices represent the elements of the sets S and T is the digraph of a function from S to T. There must be one and only one arrow leaving each vertex of the digraph representing an element of S. The fact that there is one arrow means that $f(x)$ is defined for each x in S. The fact that there is only one arrow means that each x in S is related to exactly one element of T. (Note that these remarks hold as well if we have a function from S to S and draw only one set of vertices representing the elements of S.) For further discussion of functions and digraphs see Sections A.1.1 and Subsection A.1.2 of Appendix A.

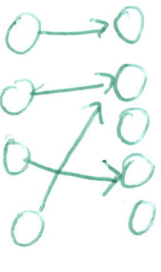

○ **Problem 23.** Draw the digraph of the function from the set {Alice, Bob, Dawn, Bill} to the set {A, B, C, D, E} given by

$$f(X) = \text{the first letter of the name } X.$$

● **Problem 24.** A function $f : S \to T$ is called an **onto function** or **surjection** if each element of T is $f(x)$ for some $x \in S$. Choose a set S and a set T so that you can draw the digraph of a function from S to T that is one-to-one but not onto, and draw the digraph of such a function.

○ **Problem 25.** Choose a set S and a set T so that you can draw the digraph of a function from S to T that is onto but not one-to-one, and draw the digraph of such a function.

● **Problem 26.** Digraphs of functions help us visualize the ideas of one-to-one functions and onto functions.

   (a) What does the digraph of a one-to-one function (injection) from a finite set X to a finite set Y look like? (Look for a test somewhat similar to the one we described for when a digraph is the digraph of a function.) (h)

   (b) What does the digraph of an onto function look like?

(c) What does the digraph of a one-to-one and onto function from a finite set S to a set T look like?

- **Problem 27.** The word *permutation* is actually used in two different ways in mathematics. A **permutation** of a set S is one-to-one from S onto S. How many permutations does an n-element set have?

Notice that there is a great deal of consistency between the use of the word permutation in Problem 27 and the use in Problem 20. If we have some way $a_1, a_2, \ldots, a_n$ of listing our set, then any other list $b_1, b_2, \ldots, b_n$ gives us the permutation of S whose rule is $f(a_i) = b_i$, and any permutation of S, say the one given by $g(a_i) = c_i$ gives us a list $c_1, c_2, \ldots, c_n$ of S. Thus there is really very little difference between the idea of a permutation of S and an n-element permutation of S when n is the size of S.

1.2.3 The bijection principle

Another name for a one-to-one and onto function is **bijection**. The digraphs marked (a), (b), and (e) in Figure 1.2.3 are digraphs of bijections. The description in Problem 26.c of the digraph of a bijection from X to Y illustrates one of the fundamental principles of combinatorial mathematics, the **bijection principle**:

> Two sets have the same size if and only if there is a bijection between them.

It is surprising how this innocent sounding principle guides us into finding insight into some otherwise very complicated proofs.

1.2.4 Counting subsets of a set

Problem 28. The **binary representation** of a number m is a list, or string, $a_1 a_2 \ldots a_k$ of zeros and ones such that $m = a_1 2^{k-1} + a_2 2^{k-2} + \cdots + a_k 2^0$. Describe a bijection between the binary representations of the integers between 0 and $2^n - 1$ and the subsets of an n-element set. What does this tell you about the number of subsets of an n-element set? (h)

Notice that the first question in Problem 8 asked you for the number of ways to choose a three element subset from a 12 element subset. You may have seen a notation like $\binom{n}{k}$, $C(n,k)$, or $_nC_k$ which stands for the number of ways to choose a k-element subset from an n-element set. The number $\binom{n}{k}$ is read as "n choose k" and is called a **binomial coefficient** for reasons we will see later on. Another frequently used way to read the binomial coefficient notation is "the number of combinations of n things taken k at a time." You are going to be asked to construct two bijections that relate to these numbers and figure out what famous formula they prove. We are going to think about subsets of the n-element set $[n] = \{1, 2, 3, \ldots, n\}$. As an example, the set of two-element subsets of $[4]$ is

$$\{\{1,2\}, \{1,3\}, \{1,4\}, \{2,3\}, \{2,4\}, \{3,4\}\}.$$

This example tells us that $\binom{4}{2} = 6$.

- **Problem 29.** Let C be the set of k-element subsets of $[n]$ that contain the number n, and let D be the set of k-element subsets of $[n]$ that don't contain n.

    (a) Let C' be the set of $(k-1)$-element subsets of $[n-1]$. Describe a bijection from C to C'. (A verbal description is fine.)

    (b) Let D' be the set of k-element subsets of $[n-1] = \{1, 2, \ldots n-1\}$. Describe a bijection from D to D'. (A verbal description is fine.)

    (c) Based on the two previous parts, express the sizes of C and D in terms of binomial coefficients involving $n - 1$ instead of n.

    (d) Apply the sum principle to C and D and obtain a formula that expresses $\binom{n}{k}$ in terms of two binomial coefficients involving $n - 1$. You have just derived the Pascal Equation that is the basis for the famous Pascal's Triangle.

1.2.5 Pascal's Triangle

The Pascal Equation that you derived in Problem 29 gives us the triangle in Figure 1.2.4. This figure has the number of k-element subsets of an n-element set as the kth number over in the nth row (we call the top row the zeroth row and the beginning entry of a row the zeroth number over). You'll see that your formula doesn't say anything about $\binom{n}{k}$ if $k = 0$ or $k = n$, but otherwise it says that each entry is the sum of the two that are above it and just to the left or right.

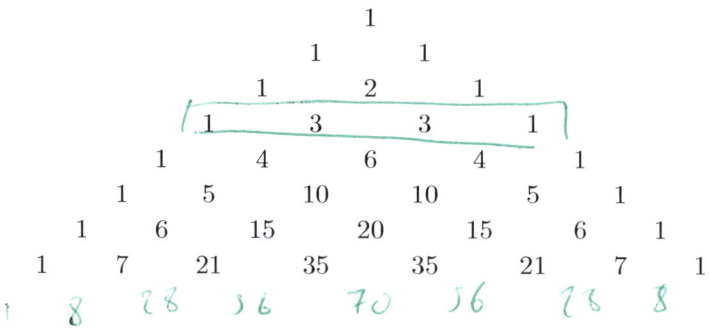

Figure 1.2.4: Pascal's Triangle

Problem 30. Just for practice, what is the next row of Pascal's triangle?

⇒ **Problem 31.** Without writing out the rows completely, write out enough of Pascal's triangle to get a numerical answer for the first question in Problem 8. (h)

It is less common to see Pascal's triangle as a right triangle, but it actually makes your formula easier to interpret. In Pascal's Right Triangle, the element in row n and column k (with the convention that the first row is row zero and the first column is column zero) is $\binom{n}{k}$. In this case your formula says each entry in a row is the sum of the one above and the one above and to the left, except for the leftmost and rightmost entries of a row, for which that doesn't make sense. Since the leftmost entry is $\binom{n}{0}$ and the rightmost entry is $\binom{n}{n}$, these entries are both one (to see why, ask yourself how many 0-element subsets and how many n-element subsets an n-element set has), and your formula then tells how to fill in the rest of the table.

	k = 0	1	2	3	4	5	6	7
n = 0	1							
1	1	1						
2	1	2	1					
3	1	3	3	1				
4	1	4	6	4	1			
5	1	5	10	10	5	1		
6	1	6	15	20	15	6	1	
7	1	7	21	35	35	21	7	1

Table 1.2.5: Pascal's Right Triangle

Seeing this right triangle leads us to ask whether there is some natural way to extend the right triangle to a rectangle. If we did have a rectangular table of binomial coefficients, counting the first row as row zero (i.e., $n = 0$) and the first column as column zero (i.e., $k = 0$), the entries we don't yet have are values of $\binom{n}{k}$ for $k > n$. But how many k-element subsets does an n-element set have if $k > n$? The answer, of course, is zero, so all the other entries we would fill in would be zero, giving us the rectangular array in Figure 1.2.6. It is straightforward to check that Pascal's equation now works for all the entries in the rectangle that have an entry above them and an entry above and to the left.

	k = 0	1	2	3	4	5	6	7
n = 0	1	0	0	0	0	0	0	0
1	1	1	0	0	0	0	0	0
2	1	2	1	0	0	0	0	0
3	1	3	3	1	0	0	0	0
4	1	4	6	4	1	0	0	0
5	1	5	10	10	5	1	0	0
6	1	6	15	20	15	6	1	0
7	1	7	21	35	35	21	7	1

Table 1.2.6: Pascal's Rectangle

⇒ **Problem 32.** Because our definition told us that $\binom{n}{k}$ is 0 when $k > n$, we got a rectangular table of numbers that satisfies the Pascal Equation.

(a) Is there any other way to define $\binom{n}{k}$ when $k > n$ in order to get a rectangular table that agrees with Pascal's Right Triangle for $k \leq n$ and satisfies the Pascal Equation? (h)

(b) Suppose we want to extend Pascal's Rectangle to the left and define $\binom{n}{-k}$ for $n \geq 0$ and $k > 0$ so that $-k < 0$. What should we put into row n and column $-k$ of Pascal's Rectangle in order for the Pascal Equation to hold true? (h)

(c) What should we put into row $-n$ and column k or column $-k$ in order for the Pascal Equation to continue to hold? Do we have any freedom of choice? (h)

Problem 33. There is yet another bijection that lets us prove that a set of size n has 2^n subsets. Namely, for each subset S of $[n] = \{1, 2, \ldots, n\}$, define a function (traditionally denoted by χ_S) as follows.[a]

$$\chi_S(i) = \begin{cases} 1 & \text{if } i \in S \\ 0 & \text{if } i \notin S \end{cases}$$

The function χ_S is called the **characteristic function** of S. Notice that the characteristic function is a function from $[n]$ to $\{0, 1\}$.

(a) For practice, consider the function $\chi_{\{1,3\}}$ for the subset $\{1, 3\}$ of the set $\{1, 2, 3, 4\}$. What are

   (i) $\chi_{\{1,3\}}(1)$?
   (ii) $\chi_{\{1,3\}}(2)$?
   (iii) $\chi_{\{1,3\}}(3)$?
   (iv) $\chi_{\{1,3\}}(4)$?

observed in Problem 34.c, how many k-
ent set have?

nula you proved in Problem 35 is symmetric in k and
s the same number for $\binom{n}{k}$ as it gives for $\binom{n}{n-k}$. Whenever
re counted by the same formula it is good for our insight
tion that demonstrates the two sets being counted have the
In fact this is a guiding principle of research in combinatorial
atics. Find a bijection that proves that $\binom{n}{k}$ equals $\binom{n}{n-k}$. (h)

• **Problem 37.** In how many ways can we pass out k (identical) ping-pong balls to n children if each child may get at most one? (h)

• **Problem 38.** In how many ways may n people sit around a round table? (Assume that when people are sitting around a round table, all that really matters is who is to each person's right. For example, if we can get one arrangement of people around the table from another by having everyone get up and move to the right one place and sit back down, we get an equivalent arrangement of people. Notice that you can get a list from a seating arrangement by marking a place at the table, and then listing the people at the table, starting at that place and moving around to the right.) There are at least two different ways of doing this problem. Try to find them both. (h)

We are now going to analyze the result of Problem 35 in more detail in order to tease out another counting principle that we can use in a wide variety of situations.

abc	acb	bac	bca	cab	cba
abd	adb	bad	bda	dab	dba
abe	aeb	bae	bea	eab	eba
acd	adc	cad	cda	dac	dca
ace	aec	cae	cea	eac	eca
ade	aed	dae	dea	ead	eda
bcd	bdc	cbd	cdb	dbc	dcb
bce	bec	cbe	ceb	ebc	ecb
bde	bed	dbe	deb	ebd	edb
cde	ced	dce	dec	ecd	edc

Table 1.2.8: The 3-element permutations of $\{a,b,c,d,e\}$ organized by which 3-element set they permute.

In Table 1.2.8 we list all three-element permutations from the 5-element set $\{a,b,c,d,e\}$. Each row consists of all 3-element permutations of some subset of $\{a,b,c,d,e\}$. Because a given k-element subset can be listed as a k-element permutation in $k!$ ways, there are $3! = 6$ permutations in each row. Because each 3-element permutation appears exactly once in the table, each row is a block of a partition of the set of 3-element permutations of $\{a,b,c,d,e\}$. Each block has size six. Each block consists of all 3-element permutations of some three-element subset of $\{a,b,c,d,e\}$. Since there are ten rows, we see that there are ten 3-element subsets of $\{a,b,c,d,e\}$. An alternate way to see this is to observe that we partitioned the set of all 60 three-element permutations of $\{a,b,c,d,e\}$ into some number q of blocks, each of size six. Thus by the product principle, $q \cdot 6 = 60$, so $q = 10$.

- **Problem 39.** Rather than restricting ourselves to $n = 5$ and $k = 3$, we can partition the set of all k-element permutations of S up into blocks. We do so by letting B_K be the set (block) of all k-element permutations of K for each k-element subset K of S. Thus as in our preceding example, each block consists of all permutations of some subset K of our n-element set. For example, the permutations of $\{a,b,c\}$ are listed in the first row of Table 1.2.8. In fact each row of that table is a block. The questions that follow are about the corresponding partition of the set of k-element permutations of S, where S and k are arbitrary.

    (a) How many permutations are there in a block? (h)

    (b) Since S has n elements, what does problem 20 tell you about the total number of k-element permutations of S?

    (c) Describe a bijection between the set of blocks of the partition and the set of k-element subsets of S. (h)

    (d) What formula does this give you for the number $\binom{n}{k}$ of k-element subsets of an n-element set? (h)

⇒ **Problem 40.** A basketball team has 12 players. However, only five players play at any given time during a game.

    (a) In how may ways may the coach choose the five players?

    (b) To be more realistic, the five players playing a game normally consist of two guards, two forwards, and one center. If there are five guards, four forwards, and three centers on the team, in how many ways can the coach choose two guards, two forwards, and one center? (h)

    (c) What if one of the centers is equally skilled at playing forward? (h)

- **Problem 41.** In Problem 38, describe a way to partition the n-element permutations of the n people into blocks so that there is a bijection between the set of blocks of the partition and the set of arrangements of the n people around a round table. What method of solution for Problem 38 does this correspond to?

- **Problem 42.** In Problems 39.d and 41, you have been using the product principle in a new way. One of the ways in which we previously stated the product principle was "If we partition a set into m blocks each of size n, then the set has size $m \cdot n$." In Problems 39.d and 41 we knew the size p of a set P of permutations of a set, and we knew we had partitioned P into some unknown number of blocks, each of a certain known size r. If we let q stand for the number of blocks, what does the product principle tell us about p, q, and r? What do we get when we solve for q?

The formula you found in the Problem 42 is so useful that we are going to single it out as another principle. The **quotient principle** says:

If we partition a set P into q blocks, each of size r, then $q = p/r$.

The quotient principle is really just a restatement of the product principle, but thinking about it as a principle in its own right often leads us to find solutions to problems. Notice that it does not always give us a formula for the number of blocks of a partition; it only works when all the blocks have the same size. In Chapter 6, we develop a way to solve problems with different block sizes in cases where there is a good deal of symmetry in the problem. (The roundness of the table was a symmetry in the problem of people at a table; the fact that we can order the sets in any order is the symmetry in the problem of counting k-element subsets.)

In Section A.2 of Appendix A we introduce the idea of an equivalence relation, see what equivalence relations have to do with partitions, and discuss the quotient principle from that point of view. While that appendix is not required for what we are doing here, if you want a more thorough discussion of the quotient principle, this would be a good time to work through that appendix.

Problem 43. In how many ways may we string n distinct beads on a necklace without a clasp? (Perhaps we make the necklace by stringing the beads on a string, and then carefully gluing the two ends of the string together so that the joint can't be seen. Assume someone can pick up the necklace, move it around in space and put it back down, giving an apparently different way of stringing the beads that is equivalent to the first.) (h)

⇒ **Problem 44.** We first gave this problem as Problem 12.a. Now we have several ways to approach the problem. A tennis club has $2n$ members. We want to pair up the members by twos for singles matches.

  (a) In how many ways may we pair up all the members of the club? Give at least two solutions different from the one you gave in Problem 12.a. (You may not have done Problem 12.a. In that case, see if you can find three solutions.) (h)

  (b) Suppose that in addition to specifying who plays whom, for each pairing we say who serves first. Now in how many ways may we specify our pairs? Try to find as many solutions as you can. (h)

• **Problem 45.** (This becomes especially relevant in Chapter 6, though it makes an important point here.) In how many ways may we attach two identical red beads and two identical blue beads to the corners of a square (with one bead per corner) free to move around in (three-dimensional) space? (h)

⇒ **Problem 46.** While the formula you proved in Problems 35 and 39.d is very useful, it doesn't give us a sense of how big the binomial coefficients are. We can get a very rough idea, for example, of the size of $\binom{2n}{n}$ by recognizing that we can write $(2n)^{\underline{n}}/n!$ as $\frac{2n}{n} \cdot \frac{2n-1}{n-1} \cdots \frac{n+1}{1}$, and each quotient is at least 2, so the product is at least 2^n. If this were an accurate estimate, it would mean the fraction of n-element subsets of a $2n$-element set would be about $2^n/2^{2n} = 1/2^n$, which becomes very small as n becomes large. However it is pretty clear the approximation will not be a very good one, because some of the terms in that product are much larger than 2. In fact, if $\binom{2n}{k}$ were the same for every k, then each would be the fraction $\frac{1}{2n+1}$ of 2^{2n}. This is much larger than the fraction $\frac{1}{2^n}$. But our intuition suggests that $\binom{2n}{n}$ is much larger than $\binom{2n}{1}$ and is likely larger than $\binom{2n}{n-1}$ so we can be sure our approximation is a bad one. For estimates like this, James Stirling developed a formula to approximate $n!$ when n is large, namely $n!$ is about $\left(\sqrt{2\pi n}\right) n^n/e^n$. In fact the ratio of $n!$ to this expression approaches 1 as n becomes infinite.[a] We write this as

$$n! \sim \sqrt{2\pi n}\frac{n^n}{e^n}.$$

We read this notation as $n!$ is asymptotic to $\sqrt{2\pi n}\frac{n^n}{e^n}$. Use Stirling's formula to show that the fraction of subsets of size n in an $2n$-element set is approximately $1/\sqrt{\pi n}$. This is a much bigger fraction than $\frac{1}{2^n}$!

> [a]Proving this takes more of a detour than is advisable here; however there is an elementary proof which you can work through in the problems of the end of Section 1 of Chapter 1 of *Introductory Combinatorics* by Kenneth P. Bogart, Harcourt Academic Press, (2000).

1.3 Some Applications of the Basic Principles

1.3.1 Lattice paths and Catalan Numbers

◦ **Problem 47.** In a part of a city, all streets run either north-south or east-west, and there are no dead ends. Suppose we are standing on a street corner. In how many ways may we walk to a corner that is four blocks north and six blocks east, using as few blocks as possible? (h)

· **Problem 48.** Problem 47 has a geometric interpretation in a coordinate plane. A **lattice path** in the plane is a "curve" made up of line segments that either go from a point (i, j) to the point $(i+1, j)$ or from a point (i, j) to the point $(i, j+1)$, where i and j are integers. (Thus lattice paths always move either up or to the right.) The length of the path is the number of such line segments.

(a) What is the length of a lattice path from $(0,0)$ to (m, n)?

(b) How many such lattice paths of that length are there? (h)

(c) How many lattice paths are there from (i, j) to (m, n), assuming i, j, m, and n are integers? (h)

· **Problem 49.** Another kind of geometric path in the plane is a **diagonal lattice path**. Such a path is a path made up of line segments that go from a point (i, j) to $(i+1, j+1)$ (this is often called an **upstep**) or $(i+1, j-1)$ (this is often called a **downstep**), again where i and j are integers. (Thus diagonal lattice paths always move towards the right but may move up or down.)

(a) Describe which points are connected to $(0,0)$ by diagonal lattice paths. (h)

(b) What is the length of a diagonal lattice path from $(0,0)$ to (m, n)?

(c) Assuming that (m, n) is such a point, how many diagonal lattice paths are there from $(0,0)$ to (m, n)? (h)

○ **Problem 50.** A school play requires a ten dollar donation per person; the donation goes into the student activity fund. Assume that each person who comes to the play pays with a ten dollar bill or a twenty dollar bill. The teacher who is collecting the money forgot to get change before the event. If there are always at least as many people who have paid with a ten as a twenty as they arrive the teacher won't have to give anyone an IOU for change. Suppose $2n$ people come to the play, and exactly half of them pay with ten dollar bills.

(a) Describe a bijection between the set of sequences of tens and twenties people give the teacher and the set of lattice paths from $(0,0)$ to (n,n).

(b) Describe a bijection between the set of sequences of tens and twenties that people give the teacher and the set of diagonal lattice paths from $(0,0)$ and $(2n,0)$.

(c) In each case, what is the geometric interpretation of a sequence that does not require the teacher to give any IOUs? (h)

⇒ • **Problem 51.** Notice that a lattice path from $(0,0)$ to (n,n) stays inside (or on the edges of) the square whose sides are the x-axis, the y-axis, the line $x = n$ and the line $y = n$. In this problem we will compute the number of lattice paths from $(0,0)$ to (n,n) that stay inside (or on the edges of) the triangle whose sides are the x-axis, the line $x = n$ and the line $y = x$. For example, in Figure 1.3.1 we show the grid of points with integer coordinates for the triangle whose sides are the x-axis, the line $x = 4$ and the line $y = x$.

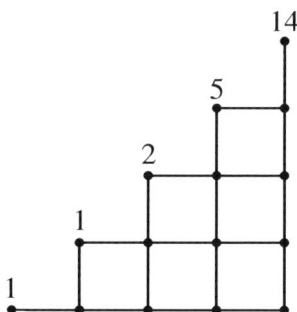

Figure 1.3.1: The lattice paths from $(0,0)$ to (i,i) for $i = 0,1,2,3,4$. The number of paths to the point (i,i) is shown just above that point.

(a) Explain why the number of lattice paths from $(0,0)$ to (n,n) that go outside the triangle is the number of lattice paths from $(0,0)$ to (n,n) that either touch or cross the line $y = x + 1$.

(b) Find a bijection between lattice paths from $(0,0)$ to (n,n) that touch (or cross) the line $y = x + 1$ and lattice paths from $(-1,1)$ to (n,n). (h)

(c) Find a formula for the number of lattice paths from $(0,0)$ to (n,n) that do not go above the line $y = x$. The number of such paths is called a **Catalan Number** and is usually denoted by C_n. (h)

⇒ **Problem 52.** Your formula for the Catalan Number can be expressed as a binomial coefficient divided by an integer. Whenever we have a formula that calls for division by an integer, an ideal combinatorial explanation of the formula is one that uses the quotient principle. The purpose of this problem is to find such an explanation using diagonal lattice paths.[a] A diagonal lattice path that never goes below the y-coordinate of its first point is called a **Dyck Path**. We will call a Dyck Path from $(0,0)$ to $(2n,0)$ a **Catalan Path** of length $2n$. Thus the number of Catalan Paths of length $2n$ is the Catalan Number C_n.

(a) If a Dyck Path has n steps (each an upstep or downstep), why do the first k steps form a Dyck Path for each nonnegative $k \leq n$?

(b) Thought of as a curve in the plane, a diagonal lattice path can have many local maxima and minima, and can have several absolute maxima and minima, that is, several highest points and several lowest points. What is the y-coordinate of an absolute minimum point of a Dyck Path starting at $(0,0)$? Explain why a Dyck Path whose rightmost absolute minimum point is its last point is a Catalan Path. (h)

(c) Let D be the set of all diagonal lattice paths from $(0,0)$ to $(2n,0)$. (Thus these paths can go below the x-axis.) Suppose we partition D by letting B_i be the set of lattice paths in D that have i upsteps (perhaps mixed with some downsteps) following the last absolute minimum. How many blocks does this partition have? Give a succinct description of the block B_0. (h)

(d) How many upsteps are in a Catalan Path?

* (e) We are going to give a bijection between the set of Catalan Paths and the block B_i for each i between 1 and n. For now, suppose the value of i, while unknown, is fixed. We take a Catalan path and break it into three pieces. The piece F (for "front") consists of all steps before the ith upstep in the Catalan path. The piece U (for "up") consists of the ith upstep. The piece B (for "back") is the portion of the path that follows the ith upstep. Thus we can think of the path as FUB. Show that the function that takes FUB to BUF is a bijection from the set of Catalan Paths onto the block B_i of the partition. (Notice that BUF can go below the x axis.) (h)

(f) Explain how you have just given another proof of the formula for the Catalan Numbers.

> [a] The result we will derive is called the Chung-Feller Theorem; this approach is based of a paper of Wen-jin Woan "Uniform Partitions of Lattice Paths and Chung-Feller Generalizations," *American Mathematics Monthly* 58 June/July 2001, p556.

1.3.2 The Binomial Theorem

○ **Problem 53.** We know that $(x+y)^2 = x^2 + 2xy + y^2$. Multiply both sides by $(x+y)$ to get a formula for $(x+y)^3$ and repeat to get a formula for $(x+y)^4$. Do you see a pattern? If so, what is it? If not, repeat the process to get a formula for $(x+y)^5$ and look back at Figure 1.2.4 to see the pattern. Conjecture a formula for $(x+y)^n$.

• **Problem 54.** When we apply the distributive law n times to $(x+y)^n$, we get a sum of terms of the form $x^i y^{n-i}$ for various values of the integer i. If it is clear to you that each term of the form $x^i y^{n-i}$ that we get comes from choosing an x from i of the $(x+y)$ factors and a y from the remaining $n-i$ of the factors and multiplying these choices together, then answer part a of the problem and skip part b. In either case, be sure to answer part c.

(a) In how many ways can we choose an x from i terms and a y from $n-i$ terms?

(b) We can take this step-by-step and consider a small case to get started.

   (i) Expand the product $(x_1 + y_1)(x_2 + y_2)(x_3 + y_3)$.

   (ii) What do you get when you substitute x for each x_i and y for each y_i?

   (iii) Now imagine expanding
   $$(x_1 + y_1)(x_2 + y_2) \cdots (x_n + y_n).$$
   Once you apply the commutative law to the individual terms you get, you will have a sum of terms of the form
   $$x_{k_1} x_{k_2} \cdots x_{k_i} \cdot y_{j_1} y_{j_2} \cdots y_{j_{n-i}}.$$
   What is the set $\{k_1, k_2, \ldots, k_i\} \cup \{j_1, j_2, \ldots, j_{n-i}\}$?

   (iv) In how many ways can you choose the set $\{k_1, k_2, \ldots, k_i\}$?

   (v) Once you have chosen this set, how many choices do you have for $\{j_1, j_2, \ldots, j_{n-i}\}$?

   (vi) If you substitute x for each x_i and y for each y_i, how many terms of the form $x^i y^{n-i}$ will you have in the expanded product
   $$(x_1 + y_1)(x_2 + y_2) \cdots (x_n + y_n) = (x+y)^n?$$

(vii) How many terms of the form $x^{n-i}y^i$ will you have?

(c) Explain how you have just proved your conjecture from Problem 53. The theorem you have proved is called the **Binomial Theorem**.

Problem 55. What is $\sum_{i=1}^{10} \binom{10}{i} 3^i$? (h)

Problem 56. What is $\binom{n}{0} - \binom{n}{1} + \binom{n}{2} - \cdots \pm \binom{n}{n}$ if n is an integer bigger than 0? (h)

Problem 57. Explain why

$$\sum_{i=0}^{m} \binom{m}{i}\binom{n}{k-i} = \binom{m+n}{k}.$$

Find two different explanations. (h)

⇒ **Problem 58.** From the symmetry of the binomial coefficients, it is not too hard to see that when n is an odd number, the number of subsets of $\{1, 2, \ldots, n\}$ of odd size equals the number of subsets of $\{1, 2, \ldots, n\}$ of even size. Is it true that when n is even the number of subsets of $\{1, 2, \ldots, n\}$ of even size equals the number of subsets of odd size? Why or why not? (h)

⇒ **Problem 59.** What is $\sum_{i=0}^{n} i\binom{n}{i}$? (Hint: think about how you might use calculus.) (h)

Notice how the proof you gave of the binomial theorem was a counting argument. It is interesting that an apparently algebraic theorem that tells us how to expand a power of a binomial is proved by an argument that amounts to counting the individual terms of the expansion. Part of the reason that combinatorial mathematics turns out to be so useful is that counting arguments often underlie important results of algebra. As the algebra becomes more sophisticated, so do the families of objects we have to count, but nonetheless we can develop a great deal of algebra on the basis of counting.

1.3.3 The pigeonhole principle

○ **Problem 60.** American coins are all marked with the year in which they were made. How many coins do you need to have in your hand to guarantee that on two (at least) of them, the date has the same last digit? (When we say "to guarantee that on two (at least) of them,..." we mean that you can find two with the same last digit. You might be able to find three with that last digit, or you might be able to find one pair with the last digit 1 and one pair with the last digit 9, or any combination of equal last digits, as long as there is at least one pair with the same last digit.)

There are many ways in which you might explain your answer to Problem 60. For example, you can partition the coins according to the last digit of their date; that is, you put all the coins with a given last digit in a block together, and put no other coins in that block; repeating until all coins are in some block. Then you have a partition of your set of coins. If no two coins have the same last digit, then each block has exactly one coin. Since there are only ten digits, there are at most ten blocks and so by the sum principle there are at most ten coins. In fact with ten coins it is possible to have no two with the same last digit, but with 11 coins some block must have at least two coins in order for the sum of the sizes of at most ten blocks to be 11. This is one explanation of why we need 11 coins in Problem 60. This kind of situation arises often in combinatorial situations, and so rather than always using the sum principle to explain our reasoning , we enunciate another principle which we can think of as yet another variant of the sum principle. The **pigeonhole principle** states that

> If we partition a set with more than n elements into n parts, then at least one part has more than one element.

The pigeonhole principle gets its name from the idea of a grid of little boxes that might be used, for example, to sort mail, or as mailboxes for a group of people in an office. The boxes in such grids are sometimes called pigeonholes in analogy with stacks of boxes used to house homing pigeons when homing pigeons were used to carry messages. People will sometimes state the principle in a more colorful way as "if we put more than n pigeons into n pigeonholes, then some pigeonhole has more than one pigeon."

Problem 61. Show that if we have a function from a set of size n to a set of size less than n, then f is not one-to-one. (h)

• **Problem 62.** Show that if S and T are finite sets of the same size, then a function f from S to T is one-to-one if and only if it is onto. (h)

- **Problem 63.** There is a **generalized pigeonhole principle** which says that if we partition a set with more than kn elements into n blocks, then at least one block has at least $k+1$ elements. Prove the generalized pigeonhole principle. (h)

Problem 64. All the powers of five end in a five, and all the powers of two are even. Show that for for some integer n, if you take the first n powers of a prime other than two or five, one must have "01" as the last two digits. (h)

⇒ · **Problem 65.** Show that in a set of six people, there is a set of at least three people who all know each other, or a set of at least three people none of whom know each other. (We assume that if person 1 knows person 2, then person 2 knows person 1.) (h)

- **Problem 66.** Draw five circles labeled Al, Sue, Don, Pam, and Jo. Find a way to draw red and green lines between people so that every pair of people is joined by a line and there is neither a triangle consisting entirely of red lines or a triangle consisting of green lines. What does Problem 65 tell you about the possibility of doing this with six people's names? What does this problem say about the conclusion of Problem 65 holding when there are five people in our set rather than six?

1.3.4 Ramsey Numbers

Problems 65–66 together show that six is the smallest number R with the property that if we have R people in a room, then there is either a set of (at least) three mutual acquaintances or a set of (at least) three mutual strangers. Another way to say the same thing is to say that six is the smallest number so that no matter how we connect 6 points in the plane (no three on a line) with red and green lines, we can find either a red triangle or a green triangle. There is a name for this property. The **Ramsey Number** $R(m, n)$ is the smallest number R so that if we have R people in a room, then there is a set of at least m mutual acquaintances or at least n mutual strangers. There is also a geometric description of Ramsey Numbers; it uses the idea of a **complete graph** on R vertices. A **complete graph** on R vertices consists of R points in the plane together with line segments (or curves) connecting each two of the R vertices.[1] The points are called **vertices** and the line segments are called **edges**. In Figure 1.3.2 we show three different ways to draw a complete graph on four vertices. We use K_n to stand for a complete graph on n vertices.

---
[1] As you may have guessed, a complete graph is a special case of something called a graph. The word graph will be defined in Subsection 2.3.1.

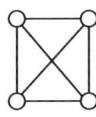 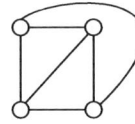 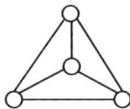

Figure 1.3.2: Three ways to draw a complete graph on four vertices

Our geometric description of $R(3,3)$ may be translated into the language of graph theory (which is the subject that includes complete graphs) by saying $R(3,3)$ is the smallest number R so that if we color the edges of a K_R with two colors, then we can find in our picture a K_3 all of whose edges have the same color. The graph theory description of $R(m,n)$ is that $R(m,n)$ is the smallest number R so that if we color the edges of a K_R with red and green, then we can find in our picture either a K_m all of whose edges are red or a K_n all of whose edges are green. Because we could have said our colors in the opposite order, we may conclude that $R(m,n) = R(n,m)$. In particular $R(n,n)$ is the smallest number R such that if we color the edges of a K_R with two colors, then our picture contains a K_n all of whose edges have the same color.

∘ **Problem 67.** Since $R(3,3) = 6$, an uneducated guess might be that $R(4,4) = 8$. Show that this is not the case. (h)

· **Problem 68.** Show that among ten people, there are either four mutual acquaintances or three mutual strangers. What does this say about $R(4,3)$? (h)

· **Problem 69.** Show that among an odd number of people there is at least one person who is an acquaintance of an even number of people and therefore also a stranger to an even number of people. (h)

· **Problem 70.** Find a way to color the edges of a K_8 with red and green so that there is no red K_4 and no green K_3. (h)

⇒ · **Problem 71.** Find $R(4,3)$. (h)

As of this writing, relatively few Ramsey Numbers are known. $R(3,n)$ is known for $n < 10$, $R(4,4) = 18$, and $R(5,4) = R(4,5) = 25$.

1.4 Supplementary Chapter Problems

⇒ **1.** Remember that we can write n as a sum of n ones. How many plus signs do we use? In how many ways may we write n as a sum of a list of k positive numbers? Such a list is called a **composition** of n into k parts.

2. In Problem 1.4.1 we defined a composition of n into k parts. What is the total number of compositions of n (into any number of parts).

· **3.** Write down a list of all 16 zero-one sequences of length four starting with 0000 in such a way that each entry differs from the previous one by changing just one digit. This is called a Gray Code. That is, a **Gray Code** for 0-1 sequences of length n is a list of the sequences so that each entry differs from the previous one in exactly one place. Can you describe how to get a Gray Code for 0-1 sequences of length five from the one you found for sequences of length 4? Can you describe how to prove that there is a Gray code for sequences of length n?

⇒ **4.** Use the idea of a Gray code from Problem 1.4.3 to prove bijectively that the number of even-sized subsets of an n-element set equals the number of odd-sized subsets of an n-element set.

⇒ **5.** A list of parentheses is said to be balanced if there are the same number of left parentheses as right, and as we count from left to right we always find at least as many left parentheses as right parentheses. For example, ((((()()))()) is balanced and ((()) and (()()))(() are not. How many balanced lists of n left and n right parentheses are there?

∗ **6.** Suppose we plan to put six distinct computers in a network as shown in Figure 1.4.1. The lines show which computers can communicate directly with which others. Consider two ways of assigning computers to the nodes of the network different if there are two computers that communicate directly in one assignment and that don't communicate directly in the other. In how many different ways can we assign computers to the network?

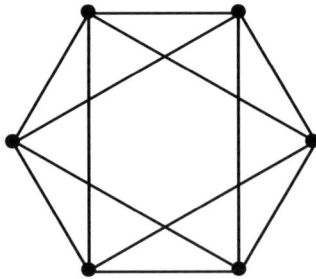

Figure 1.4.1: A computer network.

⇒ **7.** In a circular ice cream dish we are going to put four distinct scoops of ice cream chosen from among twelve flavors. Assuming we place four scoops of the same size as if they were at the corners of a square, and recognizing that moving the dish doesn't change the way in which we have put the ice cream into the dish, in how many ways may we choose the ice cream and put it into the dish?

⇒ 8. In as many ways as you can, show that $\binom{n}{k}\binom{n-k}{m} = \binom{n}{m}\binom{n-m}{k}$.

⇒ 9. A tennis club has $4n$ members. To specify a doubles match, we choose two teams of two people. In how many ways may we arrange the members into doubles matches so that each player is in one doubles match? In how many ways may we do it if we specify in addition who serves first on each team?

10. A town has n streetlights running along the north side of main street. The poles on which they are mounted need to be painted so that they do not rust. In how many ways may they be painted with red, white, blue, and green if an even number of them are to be painted green?

* 11. We have n identical ping-pong balls. In how many ways may we paint them red, white, blue, and green?

* 12. We have n identical ping-pong balls. In how many ways may we paint them red, white, blue, and green if we use green paint on an even number of them?

Chapter 2

Applications of Induction and Recursion in Combinatorics and Graph Theory

2.1 Some Examples of Mathematical Induction

If you are unfamiliar with the Principle of Mathematical Induction, you should read Appendix B (a portion of which is repeated here).

2.1.1 Mathematical induction

The **principle of mathematical induction** states that

In order to prove a statement about an integer n, if we can

1. Prove the statement when $n = b$, for some fixed integer b
2. Show that the truth of the statement for $n = k-1$ implies the truth of the statement for $n = k$ whenever $k > b$,

then we can conclude the statement is true for all integers $n \geq b$.

As an example, let us give yet another proof that a set with n elements has 2^n subsets. This proof uses essentially the same bijections we used in proving the Pascal Equation. The statement we wish to prove is the statement that "A set of size n has 2^n subsets."

Our statement is true when $n = 0$, because a set of size 0 is the empty set and the empty set has $1 = 2^0$ subsets. (This step of our proof is called a **base step**.) Now suppose that $k > 0$ and every set with $k-1$ elements has 2^{k-1} subsets. Suppose $S = \{a_1, a_2, \ldots a_k\}$ is a set with k elements. We partition the subsets of S into two blocks. Block B_1 consists of the subsets that do not contain a_n and block B_2 consists of the subsets that do contain a_n. Each set in B_1 is a subset of $\{a_1, a_2, \ldots a_{k-1}\}$, and each subset of $\{a_1, a_2, \ldots a_{k-1}\}$ is in B_1. Thus B_1 is the set of all subsets of

$\{a_1, a_2, \ldots a_{k-1}\}$. Therefore by our assumption in the first sentence of this paragraph, the size of B_1 is 2^{k-1}. Consider the function from B_2 to B_1 which takes a subset of S including a_k and removes a_k from it. This function is defined on B_2, because every set in B_2 contains a_k. This function is onto, because if T is a set in B_1, then $T \cup \{a_k\}$ is a set in B_2 which the function sends to T. This function is one-to-one because if V and W are two different sets in B_2, then removing a_k from them gives two different sets in B_1. Thus we have a bijection between B_1 and B_2, so B_1 and B_2 have the same size. Therefore by the sum principle the size of $B_1 \cup B_2$ is $2^{k-1} + 2^{k-1} = 2^k$. Therefore S has 2^k subsets. This shows that if a set of size $k-1$ has 2^{k-1} subsets, then a set of size k has 2^k subsets. Therefore by the principle of mathematical induction, a set of size n has 2^n subsets for every nonnegative integer n.

The first sentence of the last paragraph is called the **inductive hypothesis**. In an inductive proof we always make an inductive hypothesis as part of proving that the truth of our statement when $n = k - 1$ implies the truth of our statement when $n = k$. The last paragraph itself is called the **inductive step** of our proof. In an inductive step we derive the statement for $n = k$ from the statement for $n = k - 1$, thus proving that the truth of our statement when $n = k - 1$ implies the truth of our statement when $n = k$. The last sentence in the last paragraph is called the **inductive conclusion**. All inductive proofs should have a base step, an inductive hypothesis, an inductive step, and an inductive conclusion.

There are a couple details worth noticing. First, in this problem, our base step was the case $n = 0$, or in other words, we had $b = 0$. However, in other proofs, b could be any integer, positive, negative, or 0. Second, our proof that the truth of our statement for $n = k - 1$ implies the truth of our statement for $n = k$ required that k be at least 1, so that there would be an element a_k we could take away in order to describe our bijection. However, condition (2) of the principle of mathematical induction only requires that we be able to prove the implication for $k > 0$, so we were allowed to assume $k > 0$.

2.1.1.1 Strong Mathematical Induction

One way of looking at the principle of mathematical induction is that it tells us that if we know the "first" case of a theorem and we can derive each other case of the theorem from a smaller case, then the theorem is true in all cases. However the particular way in which we stated the theorem is rather restrictive in that it requires us to derive each case from the immediately preceding case. This restriction is not necessary, and removing it leads us to a more general statement of the principal of mathematical induction which people often call the **strong principle of mathematical induction**. It states:

In order to prove a statement about an integer n if we can

1. prove our statement when $n = b$ and
2. prove that the statements we get with $n = b, n = b+1, \ldots n = k-1$ imply the statement with $n = k$,

then our statement is true for all integers $n \geq b$.

You will find some explicit examples of the use of the strong principle of mathematical induction in Appendix B and will find some uses for it in this chapter.

2.1.2 Binomial Coefficients and the Binomial Theorem

- **Problem 72.** When we studied the Pascal Equation and subsets in Chapter 1, it may have appeared that there is no connection between the Pascal relation $\binom{n}{k} = \binom{n-1}{k-1} + \binom{n-1}{k}$ and the formula $\binom{n}{k} = \frac{n!}{k!(n-k)!}$. Of course you probably realize you can prove the Pascal relation by substituting the values the formula gives you into the right-hand side of the equation and simplifying to give you the left hand side. In fact, from the Pascal Relation and the facts that $\binom{n}{0} = 1$ and $\binom{n}{n} = 1$, you can actually prove the formula for $\binom{n}{k}$ by induction on n. Do so. (h)

⇒ **Problem 73.** Use the fact that $(x+y)^n = (x+y)(x+y)^{n-1}$ to give an inductive proof of the binomial theorem. (h)

Problem 74. Suppose that f is a function defined on the nonnegative integers such that $f(0) = 3$ and $f(n) = 2f(n-1)$. Find a formula for $f(n)$ and prove your formula is correct.

Problem 75. Prove the conjecture in Part 13.b for an arbitrary positive integer m without appealing to the general product principle. (h)

2.1.3 Inductive definition

You may have seen $n!$ described by the two equations $0! = 1$ and $n! = n(n-1)!$ for $n > 0$. By the principle of mathematical induction we know that this pair of equations defines $n!$ for all nonnegative numbers n. For this reason we call such a definition an **inductive definition**. An inductive definition is sometimes called a **recursive definition**. Often we can get very easy proofs of useful facts by using inductive definitions.

⇒ **Problem 76.** An inductive definition of a^n for nonnegative n is given by $a^0 = 1$ and $a^n = aa^{n-1}$. (Notice the similarity to the inductive definition of $n!$.) We remarked above that inductive definitions often give us easy proofs of useful facts. Here we apply this inductive definition to prove two useful facts about exponents that you have been using almost since you learned the meaning of exponents.

(a) Use this definition to prove the rule of exponents $a^{m+n} = a^m a^n$ for nonnegative m and n. (h)

(b) Use this definition to prove the rule of exponents $a^{mn} = (a^m)^n$. (h)

+ **Problem 77.** Suppose that f is a function on the nonnegative integers such that $f(0) = 0$ and $f(n) = n + f(n-1)$. Prove that $f(n) = n(n+1)/2$. Notice that this gives a third proof that $1 + 2 + \cdots + n = n(n+1)/2$, because this sum satisfies the two conditions for f. (The sum has no terms and is thus 0 when $n = 0$.)

$\Rightarrow$ **Problem 78.** Give an inductive definition of the summation notation $\sum_{i=1}^{n} a_i$. Use it and the distributive law $b(a + c) = ba + bc$ to prove the distributive law

$$b \sum_{i=1}^{n} a_i = \sum_{i=1}^{n} b a_i.$$

2.1.4 Proving the general product principle (Optional)

We stated the sum principle as

> If we have a partition of a set S, then the size of S is the sum of the sizes of the blocks of the partition.

In fact, the simplest form of the sum principle says that the size of the sum of two disjoint (finite) sets is the sum of their sizes.

Problem 79. Prove the sum principle we stated for partitions of a set from the simplest form of the sum principle. (h)

We stated the simplest form of the product principle as

> If we have a partition of a set S into m blocks, each of size n, then S has size mn.

In Problem 14 we gave a more general form of the product principle which can be stated as

> Let S be a set of functions f from $[n]$ to some set X. Suppose that
> - there are k_1 choices for $f(1)$, and
> - suppose that for each choice of $f(1), f(2), \ldots f(i-1)$, there are k_i choices for $f(i)$.
>
> Then the number of functions in the set S is $k_1 k_2 \cdots k_n$.

+ **Problem 80.** Prove the general form of the product principle from the simplest form of the product principle. (h)

2.1.5 Double Induction and Ramsey Numbers

In Section 1.3.4 we gave two different descriptions of the Ramsey number $R(m,n)$. However if you look carefully, you will see that we never showed that Ramsey numbers actually exist; we merely described what they were and showed that $R(3,3)$ and $R(3,4)$ exist by computing them directly. As long as we can show that there is some number R such that when there are R people together, there are either m mutual acquaintances or n mutual strangers, this shows that the Ramsey Number $R(m,n)$ exists, because it is the smallest such R. Mathematical induction allows us to show that one such R is $\binom{m+n-2}{m-1}$. The question is, what should we induct on, m or n? In other words, do we use the fact that with $\binom{m+n-3}{m-2}$ people in a room there are at least $m-1$ mutual acquaintances or n mutual strangers, or do we use the fact that with at least $\binom{m+n-3}{n-2}$ people in a room there are at least m mutual acquaintances or at least $n-1$ mutual strangers? It turns out that we use both. Thus we want to be able to simultaneously induct on m and n. One way to do that is to use yet another variation on the principle of mathematical induction, the **Principle of Double Mathematical Induction**. This principle (which can be derived from one of our earlier ones) states that

In order to prove a statement about integers m and n, if we can

1. Prove the statement when $m = a$ and $n = b$, for fixed integers a and b

2. Prove the statement when $m = a$ and $n > b$ and when $m > a$ and $n = b$ (for the same fixed integers a and b),

3. Show that the truth of the statement for $m = j$ and $n = k - 1$ (with $j \geq a$ and $k > b$) and the truth of the statement for $m = j - 1$ and $n = k$ (with $j > a$ and $k \geq b$) imply the truth of the statement for $m = j$ and $n = k$,

then we can conclude the statement is true for all pairs of integers $m \geq a$ and $n \geq b$.

There is a strong version of double induction, and it is actually easier to state. The principle of **strong double mathematical induction** says the following.

In order to prove a statement about integers m and n, if we can

1. Prove the statement when $m = a$ and $n = b$, for fixed integers a and b.

2. Show that the truth of the statemetn for values of m and n with $a + b \leq m + n < k$ imples the truth of the statment for $m + n = k$,

then we can conclude that the statement is true for all pairs of integers $m \geq a$ and $n \geq b$.

⇒ · **Problem 81.** Prove that $R(m,n)$ exists by proving that if there are $\binom{m+n-2}{m-1}$ people in a room, then there are either at least m mutual acquaintances or at least n mutual strangers. (h)

· **Problem 82.** Prove that $R(m,n) \leq R(m-1,n) + R(m,n-1)$. (h)

⇒ · **Problem 83.**

(a) What does the equation in Problem 82 tell us about $R(4,4)$?

∗ (b) Consider 17 people arranged in a circle such that each person is acquainted with the first, second, fourth, and eighth person to the right and the first, second, fourth, and eighth person to the left. can you find a set of four mutual acquaintances? Can you find a set of four mutual strangers? (h)

(c) What is $R(4,4)$?

Problem 84. (Optional) Prove the inequality of Problem 81 by induction on $m+n$.

Problem 85. Use Stirling's approximation (Problem 46) to convert the upper bound for $R(n,n)$ that you get from Problem 81 to a multiple of a power of an integer.

2.1.6 A bit of asymptotic combinatorics

Problem 83 gives us an upper bound on $R(n,n)$. A very clever technique due to Paul Erdös, called the "probabilistic method," will give a lower bound. Since both bounds are exponential in n, they show that $R(n,n)$ grows exponentially as n gets large. An analysis of what happens to a function of n as n gets large is usually called an **asymptotic analysis**. The **probabilistic method**, at least in its simpler forms, can be expressed in terms of averages, so one does not need to know the language of probability in order to understand it. We will apply it to Ramsey numbers in the next problem. Combined with the result of Problem 83, this problem will give us that $\sqrt{2}^n < R(n,n) < 2^{2n-2}$, so that we know that the Ramsey number $R(n,n)$ grows exponentially with n.

⇒ **Problem 86.** Suppose we have two numbers n and m. We consider all possible ways to color the edges of the complete graph K_m with two colors, say red and blue. For each coloring, we look at each n-element subset N of the vertex set M of K_m. Then N together with the edges of of K_m connecting vertices in N forms a complete graph on n vertices. This graph, which we denote by K_N, has its edges colored by the original coloring of the edges of K_m.

(a) Why is it that if there is no subset $N \subseteq M$ so that all the edges of K_N are colored the same color, then $R(n,n) > m$? (h)

(b) To apply the probabilistic method, we are going to compute the average, over all colorings of K_m, of the number of sets $N \subseteq M$ with $|N| = n$ such that K_N does have all its edges the same color. Explain why it is that if the average is less than 1, then for some coloring there is no set N such that K_N has all its edges colored the same color. Why does this mean that $R(n,n) > m$? (h)

(c) We call a K_N **monochromatic** for a coloring c of K_m if the color $c(e)$ assigned to edge e is the same for every edge e of K_N. Let us define $mono(c, N)$ to be 1 if N is monochromatic for c and to be 0 otherwise. Find a formula for the average number of monochromatic K_Ns over all colorings of K_m that involves a double sum first over all edge colorings c of K_m and then over all n-element subsets $N \subseteq M$ of $mono(c, N)$. (h)

(d) Show that your formula for the average reduces to $2\binom{m}{n} \cdot 2^{-\binom{n}{2}}$. (h)

(e) Explain why $R(n,n) > m$ if $\binom{m}{n} \leq 2^{\binom{n}{2}-1}$. (h)

* (f) Explain why $R(n,n) > \sqrt[n]{n! 2^{\binom{n}{2}-1}}$. (h)

(g) By using Stirling's formula, show that if n is large enough, then $R(n,n) > \sqrt{2^n} = \sqrt{2}^n$. (Here large enough means large enough for Stiirling's formula to be reasonable accurate.)

2.2 Recurrence Relations

Problem 87. How is the number of subsets of an n-element set related to the number of subsets of an $(n-1)$-element set? Prove that you are correct. (h)

Problem 88. Explain why it is that the number of bijections from an n-element set to an n-element set is equal to n times the number of bijections from an $(n-1)$-element subset to an $(n-1)$-element set. What does this have to do with Problem 27?

We can summarize these observations as follows. If s_n stands for the number of subsets of an n-element set, then

$$s_n = 2s_{n-1}, \tag{2.1}$$

and if b_n stands for the number of bijections from an n-element set to an n-element set, then

$$b_n = nb_{n-1}. \tag{2.2}$$

Equations (2.1) and (2.2) are examples of **recurrence equations** or **recurrence relations**. A **recurrence relation** or simply a **recurrence** is an equation that expresses the nth term of a sequence a_n in terms of values of a_i for $i < n$. Thus Equations (2.1) and (2.2) are examples of recurrences.

2.2.1 Examples of recurrence relations

Other examples of recurrences are

$$a_n = a_{n-1} + 7, \tag{2.3}$$

$$a_n = 3a_{n-1} + 2^n, \tag{2.4}$$

$$a_n = a_{n-1} + 3a_{n-2}, \text{ and} \tag{2.5}$$

$$a_n = a_1 a_{n-1} + a_2 a_{n-2} + \cdots + a_{n-1} a_1. \tag{2.6}$$

A **solution** to a recurrence relation is a sequence that satisfies the recurrence relation. Thus a solution to Recurrence (2.1) is $s_n = 2^n$. Note that $s_n = 17 \cdot 2^n$ and $s_n = -13 \cdot 2^n$ are also solutions to Recurrence (2.1). What this shows is that a recurrence can have infinitely many solutions. In a given problem, there is generally one solution that is of interest to us. For example, if we are interested in the number of subsets of a set, then the solution to Recurrence (2.1) that we care about is $s_n = 2^n$. Notice this is the only solution we have mentioned that satisfies $s_0 = 1$.

Problem 89. Show that there is only one solution to Recurrence (2.1) that satisfies $s_0 = 1$.

Problem 90. A first-order recurrence relation is one which expresses a_n in terms of a_{n-1} and other functions of n, but which does not include any of the terms a_i for $i < n - 1$ in the equation.

(a) Which of the recurrences (2.1) through (2.6) are first order recurrences?

(b) Show that there is one and only one sequence a_n that is defined for every nonnegative integer n, satisfies a given first order recurrence, and satisfies $a_0 = a$ for some fixed constant a. (h)

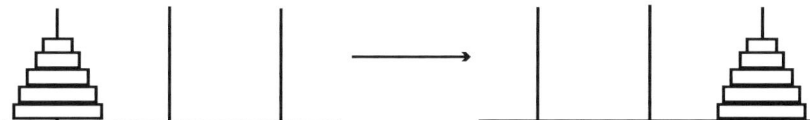

Figure 2.2.1: The Towers of Hanoi Puzzle

⇒ **Problem 91.** The "Towers of Hanoi" puzzle has three rods rising from a rectangular base with n rings of different sizes stacked in decreasing order of size on one rod. A legal move consists of moving a ring from one rod to another so that it does not land on top of a smaller ring. If m_n is the number of moves required to move all the rings from the initial rod to another rod that you choose, give a recurrence for m_n. (Hint: suppose you already knew the number of moves needed to solve the puzzle with $n-1$ rings.) (h)

⇒ **Problem 92.** We draw n mutually intersecting circles in the plane so that each one crosses each other one exactly twice and no three intersect in the same point. (As examples, think of Venn diagrams with two or three mutually intersecting sets.) Find a recurrence for the number r_n of regions into which the plane is divided by n circles. (One circle divides the plane into two regions, the inside and the outside.) Find the number of regions with n circles. For what values of n can you draw a Venn diagram showing all the possible intersections of n sets using circles to represent each of the sets? (h)

2.2.2 Arithmetic Series (optional)

Problem 93. A child puts away two dollars from her allowance each week. If she starts with twenty dollars, give a recurrence for the amount a_n of money she has after n weeks and find out how much money she has at the end of n weeks.

Problem 94. A sequence that satisfies a recurrence of the form $a_n = a_{n-1} + c$ is called an **arithmetic progression**. Find a formula in terms of the initial value a_0 and the common difference c for the term a_n in an arithmetic progression and prove you are right.

Problem 95. A person who is earning \$50,000 per year gets a raise of \$3000 a year for n years in a row. Find a recurrence for the amount a_n of money the person earns over $n+1$ years. What is the total amount of money that the person earns over a period of $n+1$ years? (In $n+1$ years, there are n raises.)

Problem 96. An **arithmetic series** is a sequence s_n equal to the sum of the terms a_0 through a_n of of an arithmetic progression. Find a recurrence for the sum s_n of an arithmetic progression with initial value a_0 and common difference c (using the language of Problem 94). Find a formula for general term s_n of an arithmetic series.

2.2.3 First order linear recurrences

Recurrences such as those in Equations (2.1) through (2.5) are called **linear recurrences**, as are the recurrences of Problems 91 and Problem 92. A **linear recurrence** is one in which a_n is expressed as a sum of functions of n times values of (some of the terms) a_i for $i < n$ plus (perhaps) another function (called the **driving function**) of n. A linear equation is called **homogeneous** if the driving function is zero (or, in other words, there is no driving function). It is called a **constant coefficient linear recurrence** if the functions that are multiplied by the a_i terms are all constants (but the driving function need not be constant).

Problem 97. Classify the recurrences in Equations (2.1) through (2.5) and Problems 91 and Problem 92 according to whether or not they are constant coefficient, and whether or not they are homogeneous.

- **Problem 98.** As you can see from Problem 97 some interesting sequences satisfy first order linear recurrences, including many that have constant coefficients, have constant driving term, or are homogeneous. Find a formula in terms of b, d, a_0 and n for the general term a_n of a sequence that satisfies a constant coefficient first order linear recurrence $a_n = ba_{n-1} + d$ and prove you are correct. If your formula involves a summation, try to replace the summation by a more compact expression. (h)

2.2.4 Geometric Series

A sequence that satisfies a recurrence of the form $a_n = ba_{n-1}$ is called a **geometric progression**. Thus the sequence satisfying Equation (2.1), the recurrence for the number of subsets of an n-element set, is an example of a geometric progression. From your solution to Problem 98, a geometric progression has the form $a_n = a_0 b^n$. In your solution to Problem 98 you may have had to deal with the sum of a geometric progression in just slightly different notation, namely $\sum_{i=0}^{n-1} db^i$. A sum of this form is called a **(finite) geometric series**.

Problem 99. Do this problem only if your final answer (so far) to Problem 98 contained the sum $\sum_{i=0}^{n-1} db^i$.

(a) Expand $(1-x)(1+x)$. Expand $(1-x)(1+x+x^2)$. Expand $(1-x)(1+x+x^2+x^3)$.

(b) What do you expect $(1-b)\sum_{i=0}^{n-1} db^i$ to be? What formula for $\sum_{i=0}^{n-1} db^i$ does this give you? Prove that you are correct.

In Problem 98 and perhaps 99 you proved an important theorem.

Theorem 2.2.2. *If $b \neq 1$ and $a_n = ba_{n-1} + d$, then $a_n = a_0 b^n + d\dfrac{1-b^n}{1-b}$. If $b = 1$, then, $a_n = a_0 + nd$*

Corollary 2.2.3. *If $b \neq 1$, then $\displaystyle\sum_{i=0}^{n-1} b^i = \dfrac{1-b^n}{1-b}$. If $b = 1$, $\displaystyle\sum_{i=0}^{n-1} b^i = n$.*

2.3 Graphs and Trees

2.3.1 Undirected graphs

In Section 1.3.4 we introduced the idea of a directed graph. Graphs consist of vertices and edges. We describe vertices and edges in much the same way as we describe points and lines in geometry: we don't really say what vertices and edges are, but we say what they do. We just don't have a complicated axiom system the way we do in geometry. A **graph** consists of a set V called a vertex set and a set E called an edge set. Each member of V is called a **vertex** and each member of E is called an **edge**. Associated with each edge are two (not necessarily different) vertices called its endpoints. We draw pictures of graphs by drawing points to represent the vertices and line segments (curved if we choose) whose endpoints are at vertices to represent the edges. In Figure 2.3.1 we show three pictures of graphs.

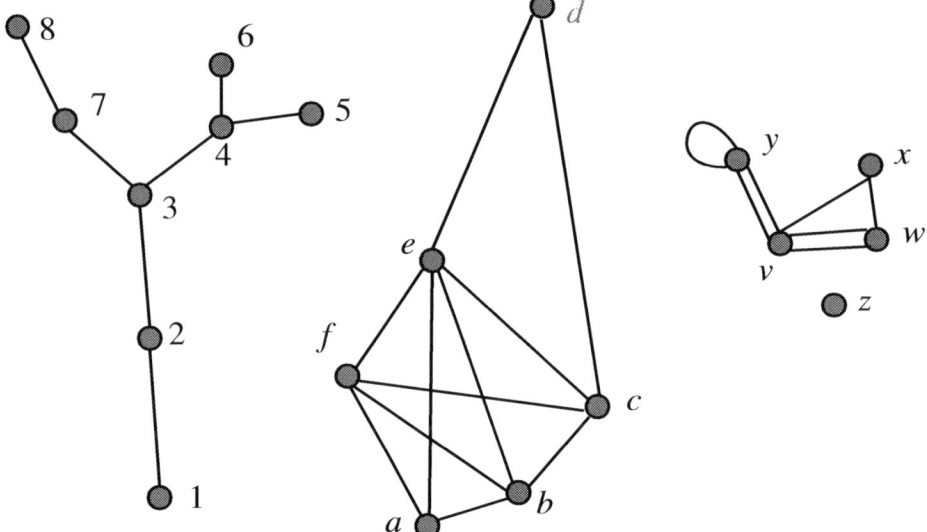

Figure 2.3.1: Three different graphs

Each gray circle in the figure represents a vertex; each line segment represents an edge. You will note that we labelled the vertices; these labels are names we chose to give the vertices. We can choose names or not as we please. The third graph also shows that it is possible to have an edge that connects a vertex (like the one labelled y) to itself or it is possible to have two or more edges (like those between vertices v and y) between two vertices. The **degree** of a vertex is the number of times it appears as the endpoint of edges; thus the degree of y in the third graph in the figure is four.

○ **Problem 100.** In the graph on the left in Figure 2.3.1, what is the degree of each vertex?

○ **Problem 101.** For each graph in Figure 2.3.1 is the number of vertices of odd degree even or odd?

⇒ · **Problem 102.** The sum of the degrees of the vertices of a (finite) graph is related in a natural way to the number of edges.

  (a) What is the relationship? (h)

  (b) Find a proof that what you say is correct that uses induction on the number of edges. Hint: To make your inductive step, think about what happens to a graph if you delete an edge. (h)

(c) Find a proof that what you say is correct that uses induction on the number of vertices.

  (d) Find a proof that what you say is correct that does not use induction. (h)

- **Problem 103.** What can you say about the number of vertices of odd degree in a graph? (h)

2.3.2 Walks and paths in graphs

A **walk** in a graph is an alternating sequence $v_0 e_1 v_1 \ldots e_i v_i$ of vertices and edges such that edge e_i connects vertices v_{i-1} and v_i. A graph is called connected if, for any pair of vertices, there is a walk starting at one and ending at the other.

Problem 104. Which of the graphs in Figure 2.3.1 is connected?

◦ **Problem 105.** A **path** in a graph is a walk with no repeated vertices. Find the longest path you can in the third graph of Figure 2.3.1.

◦ **Problem 106.** A **cycle** in a graph is a walk whose first and last vertex are equal but which has no other repeated vertices. Which graphs in Figure 2.3.1 have cycles? What is the largest number of edges in a cycle in the second graph in Figure 2.3.1? What is the smallest number of edges in a cycle in the third graph in Figure 2.3.1?

◦ **Problem 107.** A connected graph with no cycles is called a **tree**. Which graphs, if any, in Figure 2.3.1 are trees?

2.3.3 Counting vertices, edges, and paths in trees

⇒ • **Problem 108.** Draw some trees and on the basis of your examples, make a conjecture about the relationship between the number of vertices and edges in a tree. Prove your conjecture. (Hint: what happens if you choose an edge and delete it, but not its endpoints?) (h)

- **Problem 109.** What is the minimum number of vertices of degree one in a finite tree? What is it if the number of vertices is bigger than one? Prove that you are correct. (h)

⇒ · **Problem 110.** In a tree, given two vertices, how many paths can you find between them? Prove that you are correct.

⇒ ∗ **Problem 111.** How many trees are there on the vertex set $\{1,2\}$? On the vertex set $\{1,2,3\}$? When we label the vertices of our tree, we consider the tree which has edges between vertices 1 and 2 and between vertices 2 and 3 different from the tree that has edges between vertices 1 and 3 and between 2 and 3. See Figure 2.3.2.

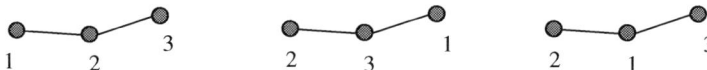

Figure 2.3.2: The three labelled trees on three vertices

How many (labelled) trees are there on four vertices? You don't have a lot of data to guess from, but try to guess a formula for the number of labelled trees with vertex set $\{1,2,\cdots,n\}$. (h)

We are now going to introduce a method to prove the formula you guessed. Given a tree with two or more vertices, labelled with positive integers, we define a sequence $b_1, b_2, \ldots$ of integers inductively as follows: If the tree has two vertices, the sequence consists of one entry, namely the label of the vertex with the larger label. Otherwise, let a_1 be the lowest numbered vertex of degree 1 in the tree. Let b_1 be the label of the unique vertex in the tree adjacent to a_1 and write down b_1. For example, in the first graph in Figure 2.3.1, a_1 is 1 and b_1 is 2. Given a_1 through a_{i-1}, let a_i be the lowest numbered vertex of degree 1 in the tree you get by deleting a_1 through a_{i-1} and let b_i be the unique vertex in this new tree adjacent to a_i. For example, in the first graph in Figure 2.3.1, $a_2 = 2$ and $b_2 = 3$. Then $a_3 = 5$ and $b_3 = 4$. We use b to stand for the sequence of b_is we get in this way. In the tree (the first graph) in Figure 2.3.1, the sequence b is 2344378. (If you are unfamiliar with inductive (recursive) definition, you might want to write down some other labelled trees on eight vertices and construct the sequence of b_is.)

Problem 112.

(a) How long will the sequence of b_is be if it is computed from a tree with n vertices (labelled with 1 through n)?

(b) What can you say about the last member of the sequence of b_is? (h)

(c) Can you tell from the sequence of b_is what a_1 is? (h)

(d) Find a bijection between labelled trees and something you can "count" that will tell you how many labelled trees there are on n labelled vertices. (h)

The sequence $b_1, b_2, \ldots, b_{n-2}$ in Problem 111 is called a **Prüfer coding** or **Prüfer code** for the tree. There is a good bit of interesting information encoded into the Prüfer code for a tree.

Problem 113. What can you say about the vertices of degree one from the Prüfer code for a tree labeled with the integers from 1 to b? (h)

Problem 114. What can you say about the Prüfer code for a tree with exactly two vertices of degree 1? (and perhaps some vertices with other degrees as well)? Does this characterize such trees?

⇒ **Problem 115.** What can you determine about the degree of the vertex labelled i from the Prüfer code of the tree? (h)

⇒ **Problem 116.** What is the number of (labelled) trees on n vertices with three vertices of degree 1? (Assume they are labelled with the integers 1 through n.) This problem will appear again in the next chapter after some material that will make it easier. (h)

2.3.4 Spanning trees

Many of the applications of trees arise from trying to find an efficient way to connect all the vertices of a graph. For example, in a telephone network, at any given time we have a certain number of wires (or microwave channels, or cellular channels) available for use. These wires or channels go from a specific place to a specific place. Thus the wires or channels may be thought of as edges of a graph and the places the wires connect may be thought of as vertices of that graph. A tree whose edges are some of the edges of a graph G and whose vertices are all of the vertices of the graph G is called a **spanning tree** of G. A spanning tree for a telephone network will give us a way to route calls between any two vertices in the network. In Figure 2.3.3 we show a graph and all its spanning trees.

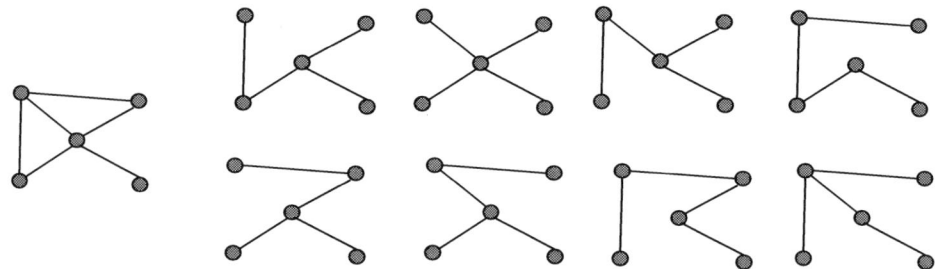

Figure 2.3.3: A graph and all its spanning trees.

> **Problem 117.** Show that every connected graph has a spanning tree. It is possible to find a proof that starts with the graph and works "down" towards the spanning tree and to find a proof that starts with just the vertices and works "up" towards the spanning tree. Can you find both kinds of proof?

2.3.5 Minimum cost spanning trees

Our motivation for talking about spanning trees was the idea of finding a minimum number of edges we need to connect all the edges of a communication network together. In many cases edges of a communication network come with costs associated with them. For example, one cell-phone operator charges another one when a customer of the first uses an antenna of the other. Suppose a company has offices in a number of cities and wants to put together a communication network connecting its various locations with high-speed computer communications, but to do so at minimum cost. Then it wants to take a graph whose vertices are the cities in which it has offices and whose edges represent possible communications lines between the cities. Of course there will not necessarily be lines between each pair of cities, and the company will not want to pay for a line connecting city i and city j if it can already connect them indirectly by using other lines it has chosen. Thus it will want to choose a spanning tree of minimum cost among all spanning trees of the communications graph. For reasons of this application, if we have a graph with numbers assigned to its edges, the sum of the numbers on the edges of a spanning tree of G will be called the **cost** of the spanning tree.

⇒ **Problem 118.** Describe a method (or better, two methods different in at least one aspect) for finding a spanning tree of minimum cost in a graph whose edges are labelled with costs, the cost on an edge being the cost for including that edge in a spanning tree. Prove that your method(s) work. (h)

The method you used in Problem 118 is called a **greedy method**, because each time you made a choice of an edge, you chose the least costly edge available to you.

2.3.6 The deletion/contraction recurrence for spanning trees

There are two operations on graphs that we can apply to get a recurrence (though a more general kind than those we have studied for sequences) which will let us compute the number of spanning trees of a graph. The operations each apply to an edge e of a graph G. The first is called **deletion**; we *delete* the edge e from the graph by removing it from the edge set. Figure 2.3.4 shows how we can delete edges from a graph to get a spanning tree.

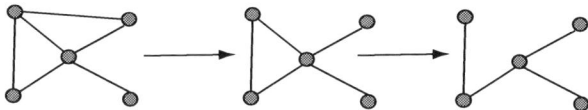

Figure 2.3.4: Deleting two appropriate edges from this graph gives a spanning tree.

The second operation is called **contraction**.

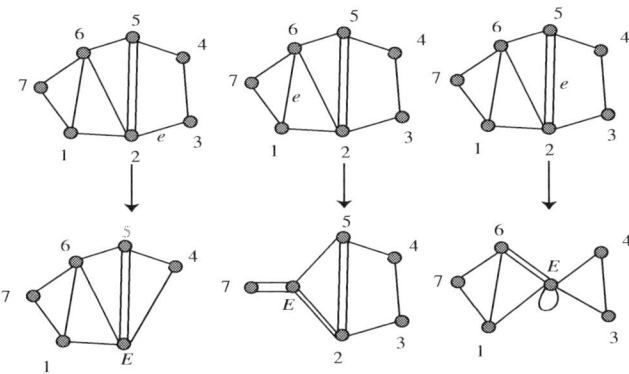

Figure 2.3.5: The results of contracting three different edges in a graph.

Contractions of three different edges in the same graph are shown in Figure 2.3.5. Intuitively, we contract an edge by shrinking it in length until its endpoints coincide; we let the rest of the graph "go along for the ride." To be more precise, we **contract** the edge e with endpoints v and w as follows:

1. remove all edges having either v or w or both as an endpoint from the edge set,
2. remove v and w from the vertex set,
3. add a new vertex E to the vertex set,
4. add an edge from E to each remaining vertex that used to be an endpoint of an edge whose other endpoint was v or w, and add an edge from E to E for any edge other than e whose endpoints were in the set $\{v, w\}$.

We use $G - e$ (read as G minus e) to stand for the result of deleting e from G, and we use G/e (read as G contract e) to stand for the result of contracting e from G.

⇒ · **Problem 119.**

(a) How do the number of spanning trees of G not containing the edge e and the number of spanning trees of G containing e relate to the number of spanning trees of $G - e$ and G/e? (h)

(b) Use $\#(G)$ to stand for the number of spanning trees of G (so that, for example, $\#(G/e)$ stands for the number of spanning trees of G/e). Find an expression for $\#(G)$ in terms of $\#(G/e)$ and $\#(G - e)$. This expression is called the **deletion-contraction recurrence**.

(c) Use the recurrence of the previous part to compute the number of spanning trees of the graph in Figure 2.3.6.

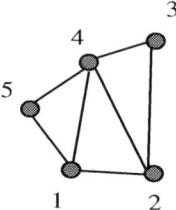

Figure 2.3.6: A graph.

2.3.7 Shortest paths in graphs

Suppose that a company has a main office in one city and regional offices in other cities. Most of the communication in the company is between the main office and the regional offices, so the company wants to find a spanning tree that minimizes not the total cost of all the edges, but rather the cost of communication between the main office and each of the regional offices. It is not clear that such a spanning tree even exists. This problem is a special case of the following. We have a connected graph with nonnegative numbers assigned to its edges. (In this situation these numbers are often called weights.) The **(weighted) length** of a path in the graph is the sum of the weights of its edges. The **distance** between two vertices is the least (weighted) length of any path between the two vertices. Given a vertex v, we would like to know the distance between v and each other vertex, and we would like to know if there is a spanning tree in G such that the length of the path in the spanning tree from v to each vertex x is the distance from v to x in G.

Problem 120. Show that the following algorithm (known as Dijkstra's algorithm) applied to a weighted graph whose vertices are labelled 1 to n gives, for each i, the distance from vertex 1 to i as $d(i)$.

1. Let $d(1) = 0$. Let $d(i) = \infty$ for all other i. Let $v(1)=1$. Let $v(j) = 0$ for all other j. For each i and j, let $w(i, j)$ be the minimum weight of an

edge between i and j, or ∞ if there are no such edges. Let $k = 1$. Let $t = 1$.

2. For each i, if $d(i) > d(k) + w(k,i)$ let $d(i) = d(k) + w(k,i)$.

3. Among those i with $v(i) = 0$, choose one with $d(i)$ a minimum, and let $k = i$. Increase t by 1. Let $v(i) = 1$.

4. Repeat the previous two steps until $t = n$

Problem 121. Is there a spanning tree such that the distance from vertex 1 to vertex i given by the algorithm in Problem 120 is the distance for vertex 1 to vertex i in the tree (using the same weights on the edges, of course)?

2.4 Supplementary Problems

1. Use the inductive definition of a^n to prove that $(ab)^n = a^n b^n$ for all nonnegative integers n.

2. Give an inductive definition of $\bigcup_{i=1}^{n} S_i$ and use it and the two set distributive law to prove the distributive law $A \cap \bigcup_{i=1}^{n} S_i = \bigcup_{i=1}^{n} A \cap S_i$.

$\Rightarrow$ 3. A hydrocarbon molecule is a molecule whose only atoms are either carbon atoms or hydrogen atoms. In a simple molecular model of a hydrocarbon, a carbon atom will bond to exactly four other atoms and hydrogen atom will bond to exactly one other atom. Such a model is shown in Figure 2.4.1. We represent a hydrocarbon compound with a graph whose vertices are labelled with C's and H's so that each C vertex has degree four and each H vertex has degree one. A hydrocarbon is called an "alkane" Common examples are methane (natural gas), butane (one version of which is shown in Figure 2.4.1)propane, hexane (ordinary gasoline), octane (to make gasoline burn more slowly), etc.

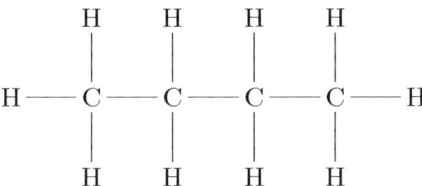

Figure 2.4.1: A model of a butane molecule

(a) How many vertices are labelled H in the graph of an alkane with exactly n vertices labelled C?

(b) An alkane is called butane if it has four carbon atoms. Why do we say one version of butane is shown in Figure 2.4.1?

4.
   (a) Give a recurrence for the number of ways to divide $2n$ people into sets of two for tennis games. (Don't worry about who serves first.)
   (b) Give a recurrence for the number of ways to divide $2n$ people into sets of two for tennis games and to determine who serves first.

$\Rightarrow$ 5. Give a recurrence for the number of ways to divide $4n$ people into sets of four for games of bridge. (Don't worry about how they sit around the bridge table or who is the first dealer.)

6. Use induction to prove your result in Supplementary Problem 1.4.2 at the end of Chapter 1.

7. Give an inductive definition of the product notation $\prod_{i=1}^{n} a_i$.

8. Using the fact that $(ab)^k = a^k b^k$, use your inductive definition of product notation in Problem 2.4.7 to prove that $\left(\prod_{i=1}^{n} a_i\right)^k = \prod_{i=1}^{n} a_i^k$.

$\Rightarrow$ 9. How many labelled trees on n vertices have exactly four vertices of degree 1?

$\Rightarrow *$ 10. The **degree sequence** of a tree is a list of the degrees of the vertices in nonincreasing order. For example the degree sequence of the first graph in Figure 2.3.3 is $(4, 3, 2, 2, 1)$. For a graph with vertices labeled 1 through n, the **ordered degree sequence** of the graph is the sequence $(d_1, d_2, \ldots, d_n)$ in which d_i is the degree of vertex i. For example, the ordered degree sqeuence of the first graph in Figure 2.3.1 is $(1, 2, 3, 3, 1, 1, 2, 1)$.
   (a) How many labelled trees are there on n vertices with ordered degree sequence $d_1, d_2, \ldots d_n$? (This problem appears again in the next chapter since some ideas in that chapter make it more straightforward.)
   (b) How many labeled trees are there on n vertices with the degree sequence in which the degree d appears i_d times?

Chapter 3

Distribution Problems

3.1 The idea of a distribution

Many of the problems we solved in Chapter 1 may be thought of as problems of distributing objects (such as pieces of fruit or ping-pong balls) to recipients (such as children). Some of the ways of viewing counting problems as distribution problems are somewhat indirect. For example, in Problem 37 you probably noticed that the number of ways to pass out k ping-pong balls to n children so that no child gets more than one is the number of ways that we may choose a k-element subset of an n-element set. We think of the children as recipients and objects we are distributing as the identical ping-pong balls, distributed so that each recipient gets at most one ball. Those children who receive an object are in our set. It is helpful to have more than one way to think of solutions to problems. In the case of distribution problems, another popular model for distributions is to think of putting balls in boxes rather than distributing objects to recipients. Passing out identical objects is modeled by putting identical balls into boxes. Passing out distinct objects is modeled by putting distinct balls into boxes.

3.1.1 The twentyfold way

When we are passing out objects to recipients, we may think of the objects as being either identical or distinct. We may also think of the recipients as being either identical (as in the case of putting fruit into plastic bags in the grocery store) or distinct (as in the case of passing fruit out to children). We may restrict the distributions to those that give at least one object to each recipient, or those that give exactly one object to each recipient, or those that give at most one object to each recipient, or we may have no such restrictions. If the objects are distinct, it may be that the order in which the objects are received is relevant (think about putting books onto the shelves in a bookcase) or that the order in which the objects are received is irrelevant (think about dropping a handful of candy into a child's trick or treat bag). If we ignore the possibility that the order in which objects are received matters, we have created $2 \cdot 2 \cdot 4 = 16$ distribution problems. In the cases where a recipient can receive more than one distinct object, we also have four more problems when the order objects are received matters. Thus we have 20 possible distribution problems.

The Twentyfold Way: A Table of Distribution Problems		
k objects and conditions on how they are received	n recipients and mathematical model for distribution	
	Distinct	Identical
1. Distinct no conditions	n^k functions	? set partitions ($\leq n$ parts)
2. Distinct Each gets at most one	$n^{\underline{k}}$ k-element permutations	1 if $k \leq n$; 0 otherwise
3. Distinct Each gets at least one	? onto functions	? set partitions (n parts)
4. Distinct Each gets exactly one	$k! = n!$ permutations	1 if $k = n$; 0 otherwise
5. Distinct, order matters	? ?	? ?
6. Distinct, order matters Each gets at least one	? ?	? ?
7. Identical no conditions	? ?	? ?
8. Identical Each gets at most one	$\binom{n}{k}$ subsets	1 if $k \leq n$; 0 otherwise
9. Identical Each gets at least one	? ?	? ?
10. Identical Each gets exactly one	1 if $k = n$; 0 otherwise	1 if $k = n$; 0 otherwise

Table 3.1.1: An incomplete table of the number of ways to distribute k objects to n recipients, with restrictions on how the objects are received

We describe these problems in Table 3.1.1. Since there are twenty possible distribution problems, we call the table the "Twentyfold Way," adapting terminology suggested by Joel Spencer for a more restricted class of distribution problems. In the first column of the table we state whether the objects are distinct (like people) or identical (like ping-pong balls) and then give any conditions on how the objects may be received. The conditions we consider are whether each recipient gets at most one object, whether each recipient gets at least one object, whether each recipient gets exactly one object, and whether the order in which the objects are received matters. In the second column we give the solution to the problem and the name of the mathematical model for this kind of distribution problem when the recipients are distinct, and in the third column we give the same information when the recipients are identical. We use question marks as the answers to problems we have not yet solved and models we have not yet studied. We give explicit answers to problems we solved in Chapter 1 and problems whose answers are immediate. The goal of this chapter is to develop methods that will allow us to fill in the table with formulas or at least quantities we know how to compute, and we will give a completed table at the end of the chapter. We will now justify the answers that are not question marks and replace some question marks with answers as we cover relevant material.

If we pass out k distinct objects (say pieces of fruit) to n distinct recipients (say children), we are saying for each object which recipient it goes to. Thus we are defining a function from the set of objects to the recipients. We saw the following theorem in Problem 13.b.

Theorem 3.1.2. *There are n^k functions from a k-element set to an n-element set.*

We proved it in Problem 13.b and in another way in Problem 75. If we pass out k distinct objects (say pieces of fruit) to n indistinguishable recipients (say identical paper bags) then we are dividing the objects up into disjoint sets; that is we are forming a partition of the objects into some number, certainly no more than the number k of objects, of parts. Later in this chapter (and again in the next chapter) we shall discuss how to compute the number of partitions of a k-element set into n parts. This explains the entries in row one of our table.

If we pass out k distinct objects to n recipients so that each gets at most one, we still determine a function, but the function must be one-to-one. The number of one-to-one functions from a k-element set to an n element set is the same as the number of one-to-one functions from the set $[k] = \{1, 2, \ldots, k\}$ to an n-element set. In Problem 20 we proved the following theorem.

Theorem 3.1.3. *If $0 \le k \le n$, then the number of k-element permutations of an n-element set is*
$$n^{\underline{k}} = n(n-1)\cdots(n-k+1) = n!/(n-k)!.$$

If $k > n$ there are no one-to-one functions from a k element set to an n element, so we define $n^{\underline{k}}$ to be zero in this case. Notice that this is what the indicated product in the middle term of our formula gives us. If we are supposed to distribute k distinct objects to n identical recipients so that each gets at most one, we cannot do so if $k > n$, so there are 0 ways to do so. On the other hand, if $k \le n$, then it doesn't matter which recipient gets which object, so there is only one way to do so. This explains the entries in row two of our table.

If we distribute k distinct objects to n distinct recipients so that each recipient gets at least one, then we are counting functions again, but this time functions from a k-element set *onto* an n-element set. At present we do not know how to compute the number of such functions, but we will discuss how to do so later in this chapter and in the next chapter. If we distribute k identical objects to n recipients, we are again simply partitioning the objects, but the condition that each recipient gets at least one means that we are partitioning the objects into exactly n blocks. Again, we will discuss how compute the number of ways of partitioning a set of k objects into n blocks later in this chapter. This explains the entries in row three of our table.

If we pass out k distinct objects to n recipients so that each gets exactly one, then $k = n$ and the function that our distribution gives us is a bijection. The number of bijections from an n-element set to an n-element set is $n!$ by Theorem 3.1.3. If we pass out k distinct objects of n identical recipients so that each gets exactly 1, then in this case it doesn't matter which recipient gets which object, so the number of ways to do so is 1 if $k = n$. If $k \neq n$, then the number of such distributions is zero. This explains the entries in row four of our table.

We now jump to row eight of our table. We saw in Problem 37 that the number of ways to pass out k identical ping-pong balls to n children is simply the number of k-element subsets of an n-element set. In Problem 39 we proved the following theorem.

Theorem 3.1.4. *If $0 \leq k \leq n$, the number of k-element subsets of an n-element set is given by*
$$\binom{n}{k} = \frac{n^{\underline{k}}}{k!} = \frac{n!}{k!(n-k)!}.$$

We define $\binom{n}{k}$ to be 0 if $k > n$, because then there are no k-element subsets of an n-element set. Notice that this is what the middle term of the formula in the theorem gives us. This explains the entries of row 8 of our table. For now we jump over row 9.

In row 10 of our table, if we are passing out k identical objects to n recipients so that each gets exactly one, it doesn't matter whether the recipients are identical or not; there is only one way to pass out the objects if $k = n$ and otherwise it is impossible to make the distribution, so there are no ways of distributing the objects. This explains the entries of row 10 of our table. Several other rows of our table can be computed using the methods of Chapter 1.

3.1.2 Ordered functions

Problem 122. Suppose we wish to place k distinct books onto the shelves of a bookcase with n shelves. For simplicity, assume for now that all of the books would fit on any of the shelves. Also, let's imagine pushing the books on a shelf as far to the left as we can, so that we are only thinking about how the books sit relative to each other, not about the exact places where we put the books. Since the books are distinct, we can think of a the first book, the second book and so on.

  (a) How many places are there where we can place the first book?

  (b) When we place the second book, if we decide to place it on the shelf that already has a book, does it matter if we place it to the left or right of the book that is already there?

  (c) How many places are there where we can place the second book? (h)

  (d) Once we have $i - 1$ books placed, if we want to place book i on a shelf that already has some books, is sliding it in to the left of all the books already there different from placing it to the right of all the books already or between two books already there?

(e) In how many ways may we place the *i*th book into the bookcase? (h)

(f) In how many ways may we place all the books?

Problem 123. Suppose we wish to place the books in Problem 122 (satisfying the assumptions we made there) so that each shelf gets at least one book. Now in how many ways may we place the books? (Hint: how can you make sure that each shelf gets at least one book before you start the process described in Problem 122?) (h)

The assignment of which books go to which shelves of a bookcase is simply a function from the books to the shelves. But a function doesn't determine which book sits to the left of which others on the shelf, and this information is part of how the books are arranged on the shelves. In other words, the order in which the shelves receive their books matters. Our function must thus assign an ordered list of books to each shelf. We will call such a function an ordered function. More precisely, an **ordered function** from a set S to a set T is a function that assigns an (ordered) list of elements of S to some, but not necessarily all, elements of T in such a way that each element of S appears on one and only one of the lists.[1] (Notice that although it is not the usual definition of a function from S to T, a function can be described as an assignment of subsets of S to some, but not necessarily all, elements of T so that each element of S is in one and only one of these subsets.) Thus the number of ways to place the books into the bookcase is the entry in the middle column of row 5 of our table. If in addition we require each shelf to get at least one book, we are discussing the entry in the middle column of row 6 of our table. An **ordered onto function** is one which assigns a list to each element of T. In Problem 122 you showed that the number of ordered functions from a k-element set to an n-element set is $\prod_{i=1}^{k}(n+i-1)$. This product occurs frequently enough that it has a name; it is called the kth **rising factorial power** of n and is denoted by $n^{\overline{k}}$. It is read as "n to the k rising." (This notation is due to Don Knuth, who also suggested the notation for falling factorial powers.) We can summarize with a theorem that adds two more formulas for the number of ordered functions.

Theorem 3.1.5. *The number of ordered functions from a k-element set to an n-element set is*

$$n^{\overline{k}} = \prod_{i=1}^{k}(n+i-1) = \frac{(n+i-1)!}{(n-1)!} = (n+k-1)^{\underline{k}}.$$

Ordered functions explain the entries in the middle column of rows 5 and 6 of our table of distribution problems.

---

[1] The phrase ordered function is not a standard one, because there is as yet no standard name for the result of an ordered distribution problem.

3.1.3 Multisets

In the middle column of row 7 of our table, we are asking for the number of ways to distribute k identical objects (say ping-pong balls) to n distinct recipients (say children).

- **Problem 124.** In how many ways may we distribute k identical books on the shelves of a bookcase with n shelves, assuming that any shelf can hold all the books? (h)

- **Problem 125.** A **multiset** chosen from a set S may be thought of as a subset with repeated elements allowed. For example the multiset of letters of the word Mississippi is $\{i, i, i, i, m, p, p, s, s, s, s\}$. To determine a multiset we must say how many times (including, perhaps, zero) each member of S appears in the multiset. The number of times an element appears is called its **multiplicity**. The size of a multiset chosen from S is the total number of times any member of S appears. For example, the size of the multiset of letters of Mississippi is 11. What is the number of multisets of size k that can be chosen from an n-element set? (h)

⇒ **Problem 126.** Your answer in the previous problem should be expressible as a binomial coefficient. Since a binomial coefficient counts subsets, find a bijection between subsets of something and multisets chosen from a set S. (h)

Problem 127. How many solutions are there in nonnegative integers to the equation $x_1 + x_2 + \cdots + x_m = r$, where m and r are constants? (h)

Problem 128. In how many ways can we distribute k identical objects to n distinct recipients so that each recipient gets at least m? (h)

Multisets explain the entry in the middle column of row 7 of our table of distribution problems.

3.1.4 Compositions of integers

- **Problem 129.** In how many ways may we put k identical books onto n shelves if each shelf must get at least one book? (h)

- **Problem 130.** A **composition** of the integer k into n parts is a list of n positive integers that add to k. How many compositions are there of an integer k into n parts? (h)

⇒ **Problem 131.** Your answer in Problem 130 can be expressed as a binomial coefficient. This means it should be possible to interpret a composition as a subset of some set. Find a bijection between compositions of k into n parts and certain subsets of some set. Explain explicitly how to get the composition from the subset and the subset from the composition. (h)

- **Problem 132.** Explain the connection between compositions of k into n parts and the problem of distributing k identical objects to n recipients so that each recipient gets at least one.

The sequence of problems you just completed should explain the entry in the middle column of row 9 of our table of distribution problems.

3.1.5 Broken permutations and Lah numbers

⇒ · **Problem 133.** In how many ways may we stack k distinct books into n identical boxes so that there is a stack in every box? The hints may suggest that you can do this problem in more than one way! (h)

We can think of stacking books into identical boxes as partitioning the books and then ordering the blocks of the partition. This turns out not to be a useful computational way of visualizing the problem because the number of ways to order the books in the various stacks depends on the sizes of the stacks and not just the number of stacks. However this way of thinking actually led to the first hint in Problem 133. Instead of dividing a set up into nonoverlapping parts, we may think of dividing a *permutation* (thought of as a list) of our k objects up into n ordered blocks. We will say that a set of ordered lists of elements of a set S is a **broken permutation** of S if each element of S is in one and only one of these lists.[2] The number of broken permutations of a k-element set with n blocks is denoted by $L(k,n)$. The number $L(k,n)$ is called a **Lah Number** and, from our solution to Problem 133, is equal to $k!\binom{k-1}{n-1}/n!$.

The Lah numbers are the solution to the question "In how many ways may we distribute k distinct objects to n identical recipients if order matters and each recipient must get at least one?" Thus they give the entry in row 6 and column 3 of our table. The entry in row 5 and column 3 of our table will be the number of

---

[2] The phrase broken permutation is not standard, because there is no standard name for the solution to this kind of distribution problem.

broken permutations with less than or equal to n parts. Thus it is a sum of Lah numbers.

We have seen that ordered functions and broken permutations explain the entries in rows 5 and 6 of our table.

In the next two sections we will give ways of computing the remaining entries.

3.2 Partitions and Stirling Numbers

We have seen how the number of partitions of a set of k objects into n blocks corresponds to the distribution of k distinct objects to n identical recipients. While there is a formula that we shall eventually learn for this number, it requires more machinery than we now have available. However there is a good method for computing this number that is similar to Pascal's equation. Now that we have studied recurrences in one variable, we will point out that Pascal's equation is in fact a *recurrence in two variables*; that is it lets us compute $\binom{n}{k}$ in terms of values of $\binom{m}{i}$ in which either $m < n$ or $i < k$ or both. It was the fact that we had such a recurrence and knew $\binom{n}{0}$ and $\binom{n}{n}$ that let us create Pascal's triangle.

3.2.1 Stirling Numbers of the second kind

We use the notation $S(k, n)$ to stand for the number of partitions of a k element set with n blocks. For historical reasons, $S(k, n)$ is called a **Stirling number of the second kind**.

Problem 134. In a partition of the set $[k]$, the number k is either in a block by itself, or it is not. How does the number of partitions of $[k]$ with n parts in which k is in a block with other elements of $[k]$ compare to the number of partitions of $[k-1]$ into n blocks? Find a two variable recurrence for $S(n, k)$, valid for n and k larger than one. (h)

Problem 135. What is $S(k, 1)$? What is $S(k, k)$? Create a table of values of $S(k, n)$ for k between 1 and 5 and n between 1 and k. This table is sometimes called **Stirling's Triangle (of the second kind)** How would you define $S(k, n)$ for the nonnegative values of k and n that are not both positive? Now for what values of k and n is your two variable recurrence valid?

Problem 136. Extend Stirling's triangle enough to allow you to answer the following question and answer it. (Don't fill in the rows all the way; the work becomes quite tedious if you do. Only fill in what you need to answer this question.) A caterer is preparing three bag lunches for hikers. The caterer has nine different sandwiches. In how many ways can these nine sandwiches be distributed into three identical lunch bags so that each bag gets at least one?

Problem 137. The question in Problem 136 naturally suggests a more realistic question; in how many ways may the caterer distribute the nine sandwich's into three identical bags so that each bag gets exactly three? Answer this question. (h)

· **Problem 138.** What is $S(k, k-1)$? (h)

• **Problem 139.** In how many ways can we partition k items into n blocks so that we have k_i blocks of size i for each i? (Notice that $\sum_{i=1}^{k} k_i = n$ and $\sum_{i=1}^{k} i k_i = k$.) The sequence $k_1, k_2, \ldots, k_n$ is called the **type vector** of the partition. (h)

+ **Problem 140.** Describe how to compute $S(k, n)$ in terms of quantities given by the formula you found in Problem 139.

⇒ **Problem 141.** Find a recurrence for the Lah numbers $L(k, n)$ similar to the one in Problem 134. (h)

· **Problem 142.** (Relevant in Appendix C.) The total number of partitions of a k-element set is denoted by $B(k)$ and is called the k-th **Bell number**. Thus $B(1) = 1$ and $B(2) = 2$.

   (a) Show, by explicitly exhibiting the partitions, that $B(3) = 5$.

   (b) Find a recurrence that expresses $B(k)$ in terms of $B(n)$ for $n < k$ and prove your formula correct in as many ways as you can. (h)

   (c) Find $B(k)$ for $k = 4, 5, 6$.

3.2.2 Stirling Numbers and onto functions

○ **Problem 143.** Given a function f from a k-element set K to an n-element set, we can define a partition of K by putting x and y in the same block of the partition if and only if $f(x) = f(y)$. How many blocks does the partition have if f is onto? How is the number of functions from a k-element set onto an n-element set related to a Stirling number? Be as precise in your answer as you can. (h)

⇒ **Problem 144.** How many labeled trees on n vertices have exactly 3 vertices of degree one? Note that this problem has appeared before in Chapter 2. (h)

• **Problem 145.** Each function from a k-element set K to an n-element set N is a function from K onto *some* subset of N. If J is a subset of N of size j, you know how to compute the number of functions that map onto J in terms of Stirling numbers. Suppose you add the number of functions mapping onto J over all possible subsets J of N. What simple value should this sum equal? Write the equation this gives you. (h)

○ **Problem 146.** In how many ways can the sandwiches of Problem 136 be placed into three distinct bags so that each bag gets at least one?

○ **Problem 147.** In how many ways can the sandwiches of Problem 137 be placed into distinct bags so that each bag gets exactly three?

• **Problem 148.** In how many ways may we label the elements of a k-element set with n distinct labels (numbered 1 through n) so that label i is used j_i times? (If we think of the labels as $y_1, y_2, \ldots, y_n$, then we can rephrase this question as follows. How many functions are there from a k-element set K to a set $N = \{y_1, y_2, \ldots y_n\}$ so that y_i is the image of j_i elements of K?) This number is called a **multinomial coefficient** and denoted by

$$\binom{k}{j_1, j_2, \ldots, j_n}.$$

(h)

Problem 149. Explain how to compute the number of functions from a k-element set K to an n-element set N by using multinomial coefficients. (h)

Problem 150. Explain how to compute the number of functions from a k-element set K onto an n-element set N by using multinomial coefficients. (h)

- **Problem 151.** What do multinomial coefficients have to do with expanding the kth power of a multinomial $x_1 + x_2 + \cdots + x_n$? This result is called the **multinomial theorem**. (h)

3.2.3 Stirling Numbers and bases for polynomials

· **Problem 152.**

(a) Find a way to express n^k in terms of $k^{\underline{j}}$ for appropriate values j. You may use Stirling numbers if they help you. (h)

(b) Notice that $x^{\underline{j}}$ makes sense for a numerical variable x (that could range over the rational numbers, the real numbers, or even the complex numbers instead of only the nonnegative integers, as we are implicitly assuming n does), just as x^j does. Find a way to express the power x^k in terms of the polynomials $x^{\underline{j}}$ for appropriate values of j and explain why your formula is correct. (h)

You showed in Problem 152 how to get each power of x in terms of the falling factorial powers $x^{\underline{j}}$. Therefore every polynomial in x is expressible in terms of a sum of numerical multiples of falling factorial powers. Using the language of linear algebra, we say that the ordinary powers of x and the falling factorial powers of x each form a basis for the "space" of polynomials, and that the numbers $S(k, n)$ are "change of basis coefficients." If you are not familiar with linear algebra, a **basis** for the **space of polynomials**[3] is a set of polynomials such that each polynomial, whether in that set or not, can be expressed in one and only one way as a sum of numerical multiples of polynomials in the set.

○ **Problem 153.** Show that every power of $x + 1$ is expressible as a sum of numerical multiples of powers of x. Now show that every power of x (and thus every polynomial in x) is a sum of numerical multiples (some of which could be negative) of powers of $x + 1$. This means that the powers of $x + 1$ are a basis for the space of polynomials as well. Describe the change of basis

---
[3] The space of polynomials is just another name for the set of all polynomials.

coefficients that we use to express the binomial powers $(x+1)^n$ in terms of the ordinary x^j explicitly. Find the change of basis coefficients we use to express the ordinary powers x^n in terms of the binomial powers $(x+1)^k$. (h)

⇒ · **Problem 154.** By multiplication, we can see that every falling factorial polynomial can be expressed as a sum of numerical multiples of powers of x. In symbols, this means that there are numbers $s(k,n)$ (notice that this s is lower case, not upper case) such that we may write $x^{\underline{k}} = \sum_{n=0}^{k} s(k,n)x^n$. These numbers $s(k,n)$ are called Stirling Numbers of the first kind. By thinking algebraically about what the formula

$$x^{\underline{k}} = x^{\underline{k-1}}(x-k+1) \qquad (3.1)$$

means, we can find a recurrence for Stirling numbers of the first kind that gives us another triangular array of numbers called Stirling's triangle of the first kind. Explain why Equation (3.1) is true and use it to derive a recurrence for $s(k,n)$ in terms of $s(k-1,n-1)$ and $s(k-1,n)$. (h)

Problem 155. Write down the rows of Stirling's triangle of the first kind for $k = 0$ to 6.

By definition, the Stirling numbers of the first kind are also change of basis coefficients. The Stirling numbers of the first and second kind are change of basis coefficients from the falling factorial powers of x to the ordinary factorial powers, and vice versa.

⇒ **Problem 156.** Explain why every rising factorial polynomial $x^{\overline{k}}$ can be expressed in terms of the falling factorial polynomials $x^{\underline{n}}$. Let $b(k,n)$ stand for the change of basis coefficients that allow us to express $x^{\overline{k}}$ in terms of the falling factorial polynomials $x^{\underline{n}}$; that is, define $b(k,n)$ by the equations

$$x^{\overline{k}} = \sum_{n=0}^{k} b(k,n) x^{\underline{n}}.$$

(a) Find a recurrence for $b(k,n)$. (h)

(b) Find a formula for $b(k,n)$ and prove the correctness of what you say in as many ways as you can. (h)

(c) Is $b(k,n)$ the same as any of the other families of numbers (binomial coefficients, Bell numbers, Stirling numbers, Lah numbers, etc.) we have studied?

(d) Say as much as you can (but say it precisely) about the change of basis coefficients for expressing $x^{\underline{k}}$ in terms of $x^{\overline{n}}$. (h)

3.3 Partitions of Integers

We have now completed all our distribution problems except for those in which both the objects and the recipients are identical. For example, we might be putting identical apples into identical paper bags. In this case all that matters is how many bags get one apple (how many recipients get one object), how many get two, how many get three, and so on. Thus for each bag we have a number, and the multiset of numbers of apples in the various bags is what determines our distribution of apples into identical bags. A multiset of positive integers that add to n is called a **partition** of n. Thus the partitions of 3 are 1+1+1, 1+2 (which is the same as 2+1) and 3. The number of partitions of k is denoted by $P(k)$; in computing the partitions of 3 we showed that $P(3) = 3$. It is traditional to use Greek letters like λ (the Greek letter λ is pronounced LAMB duh) to stand for partitons; we might write $\lambda = 1, 1, 1$, $\gamma = 2, 1$ and $\tau = 3$ to stand for the three partitions we just described. We also write $\lambda = 1^3$ as a shorthand for $\lambda = 1, 1, 1$, and we write $\lambda \dashv 3$ as a shorthand for "λ is a partition of three."

○ **Problem 157.** Find all partitions of 4 and find all partitions of 5, thereby computing $P(4)$ and $P(5)$.

3.3.1 The number of partitions of k into n parts

A **partition of the integer k into n parts** is a multiset of n positive integers that add to k. We use $P(k, n)$ to denote the number of partitions of k into n parts. Thus $P(k, n)$ is the number of ways to distribute k identical objects to n identical recipients so that each gets at least one.

○ **Problem 158.** Find $P(6, 3)$ by finding all partitions of 6 into 3 parts. What does this say about the number of ways to put six identical apples into three identical bags so that each bag has at least one apple?

3.3.2 Representations of partitions

○ **Problem 159.** How many solutions are there in the positive integers to the equation $x_1 + x_2 + x_3 = 7$ with $x_1 \geq x_2 \geq x_3$?

Problem 160. Explain the relationship between partitions of k into n parts and lists $x_1, x_2, \ldots, x_n$ of positive integers that add to k with $x_1 \geq x_2 \geq \ldots \geq x_n$. Such a representation of a partition is called a **decreasing list** representation of the partition.

○ **Problem 161.** Describe the relationship between partitions of k and lists or vectors $(x_1, x_2, \ldots, x_n)$ such that $x_1 + 2x_2 + \ldots kx_k = k$. Such a representation of a partition is called a **type vector** representation of a partition, and it is typical to leave the trailing zeros out of such a representation; for example $(2, 1)$ stands for the same partition as $(2, 1, 0, 0)$. What is the decreasing list representation for this partition, and what number does it partition?

Problem 162. How does the number of partitions of k relate to the number of partitions of $k + 1$ whose smallest part is one? (h)

When we write a partition as $\lambda = \lambda_1, \lambda_2, \ldots, \lambda_n$, it is customary to write the list of λ_is as a decreasing list. When we have a type vector $(t_1, t_2, \ldots, t_m)$ for a partition, we write either $\lambda = 1^{t_1} 2^{t_2} \cdots m^{t_m}$ or $\lambda = m^{t_m}(m-1)^{t_{m-1}} \cdots 2^{t_2} 1^{t_1}$. Henceforth we will use the second of these. When we write $\lambda = \lambda_1^{i_1} \lambda_2^{i_2} \cdots \lambda_n^{i_n}$, we will assume that $\lambda_i > \lambda_{i+1}$.

3.3.3 Ferrers and Young Diagrams and the conjugate of a partition

The decreasing list representation of partitions leads us to a handy way to visualize partitions. Given a decreasing list $(\lambda_1, \lambda_2, \ldots \lambda_n)$, we draw a figure made up of rows of dots that has λ_1 equally spaced dots in the first row, λ_2 equally spaced dots in the second row, starting out right below the beginning of the first row and so on. Equivalently, instead of dots, we may use identical squares, drawn so that a square touches each one to its immediate right or immediately below it along an edge. See Figure 3.3.1 for examples. The figure we draw with dots is called the **Ferrers diagram** of the partition; sometimes the figure with squares is also called a Ferrers diagram; sometimes it is called a **Young diagram**. At this stage it is irrelevant which name we choose and which kind of figure we draw; in more advanced work the squares are handy because we can put things like numbers or variables into them. From now on we will use squares and call the diagrams Young diagrams.

Figure 3.3.1: The Ferrers and Young diagrams of the partition (5,3,3,2)

- **Problem 163.** Draw the Young diagram of the partition (4,4,3,1,1). Describe the geometric relationship between the Young diagram of (5,3,3,2) and the Young diagram of (4,4,3,1,1). (h)

- **Problem 164.** The partition $(\lambda_1, \lambda_2, \ldots, \lambda_n)$ is called the **conjugate** of the partition $(\gamma_1, \gamma_2, \ldots, \gamma_m)$ if we obtain the Young diagram of one from the Young diagram of the other by flipping one around the line with slope -1 that extends the diagonal of the top left square. See Figure 3.3.2 for an example.

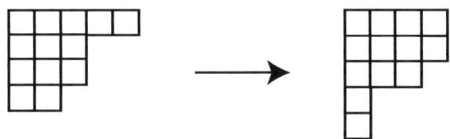

  Figure 3.3.2: The Ferrers diagram the partition (5,3,3,2) and its conjugate.

  What is the conjugate of (4,4,3,1,1)? How is the largest part of a partition related to the number of parts of its conjugate? What does this tell you about the number of partitions of a positive integer k with largest part m? (h)

⇒ **Problem 165.** A partition is called **self-conjugate** if it is equal to its conjugate. Find a relationship between the number of self-conjugate partitions of k and the number of partitions of k into distinct odd parts. (h)

Problem 166. Explain the relationship between the number of partitions of k into even parts and the number of partitions of k into parts of even multiplicity, i.e. parts which are each used an even number of times as in (3,3,3,3,2,2,1,1). (h)

⇒ **Problem 167.** Show that the number of partitions of k into four parts equals the number of partitions of $3k$ into four parts of size at most $k - 1$ (or $3k - 4$ into four parts of size at most $k - 2$ or $3k - 4$ into four parts of size at most k). (h)

Problem 168. The idea of conjugation of a partition could be defined without the geometric interpretation of a Young diagram, but it would seem far less natural without the geometric interpretation. Another idea that seems much more natural in a geometric context is this. Suppose we have a partition of k into n parts with largest part m. Then the Young diagram of the partition can fit into a rectangle that is m or more units wide (horizontally) and n or more units deep. Suppose we place the Young diagram of our partition in the top left-hand corner of an m' unit wide and n' unit deep rectangle with $m' \geq m$ and $n' \geq n$, as in Figure 3.3.3.

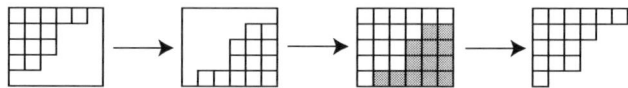

Figure 3.3.3: To complement the partition $(5, 3, 3, 2)$ in a 6 by 5 rectangle: enclose it in the rectangle, rotate, and cut out the original Young diagram.

(a) Why can we interpret the part of the rectangle not occupied by our Young diagram, rotated in the plane, as the Young diagram of another partition? This is called the **complement** of our partition in the rectangle.

(b) What integer is being partitioned by the complement?

(c) What conditions on m' and n' guarantee that the complement has the same number of parts as the original one? (h)

(d) What conditions on m' and n' guarantee that the complement has the same largest part as the original one? (h)

(e) Is it possible for the complement to have both the same number of parts and the same largest part as the original one?

(f) If we complement a partition in an m' by n' box and then complement that partition in an m' by n' box again, do we get the same partition that we started with?

⇒ **Problem 169.** Suppose we take a partition of k into n parts with largest part m, complement it in the smallest rectangle it will fit into, complement the result in the smallest rectangle it will fit into, and continue the process until we get the partition 1 of one into one part. What can you say about the partition with which we started? (h)

Problem 170. Show that $P(k, n)$ is at least $\frac{1}{n!}\binom{k-1}{n-1}$. (h)

With the binomial coefficients, with Stirling numbers of the second kind, and with the Lah numbers, we were able to find a recurrence by asking what happens to our subset, partition, or broken permutation of a set S of numbers if we remove the largest element of S. Thus it is natural to look for a recurrence to count the number of partitions of k into n parts by doing something similar. Unfortunately, since we are counting distributions in which all the objects are identical, there is no way for us to identify a largest element. However if we think geometrically, we can ask what we could remove from a Young diagram to get a Young diagram. Two natural ways to get a partition of a smaller integer from a partition of n would be to remove the top row of the Young diagram of the partition and to remove the left column of the Young diagram of the partition. These two operations correspond to removing the largest part from the partition and to subtracting 1 from each part of the partition respectively. Even though they are symmetric with respect to conjugation, they aren't symmetric with respect to the number of parts. Thus one might be much more useful than the other for finding a recurrence for the number of partitions of k into n parts.

⇒ · **Problem 171.** In this problem we will study the two operations and see which one seems more useful for getting a recurrence for $P(k, n)$.

(a) How many parts does the remaining partition have when we remove the largest part (more precisely, we reduce its multiplicity by one) from a partition of k into n parts? What can you say about the number of parts of the remaining partition if we remove one from each part? (h)

(b) If we remove the largest part from a partition, what can we say about the integer that is being partitioned by the remaining parts of the partition? If we remove one from each part of a partition of k into n parts, what integer is being partitioned by the remaining parts? (Another way to describe this is that we remove the first column from the Young diagram of the partition.) (h)

(c) The last two questions are designed to get you thinking about how we can get a bijection between the set of partitions of k into n parts and some other set of partitions that are partitions of a smaller number. These questions describe two different strategies for getting that set of partitions of a smaller number or of smaller numbers. Each strategy leads to a bijection between partitions of k into n parts and a set of partitions of a smaller number or numbers. For each strategy, use the answers to the last two questions to find and describe this set of partitions into a smaller number and a bijection between partitions of k into n parts and partitions of the smaller integer or integers into appropriate numbers of parts. (In one case the set of partitions and bijection are relatively straightforward to describe and in the other case not so easy.) (h)

(d) Find a recurrence (which need not have just two terms on the right hand side) that describes how to compute $P(k, n)$ in terms of the number of partitions of smaller integers into a smaller number of parts. (h)

(e) What is $P(k, 1)$ for a positive integer k?

(f) What is $P(k, k)$ for a positive integer k?

(g) Use your recurrence to compute a table with the values of $P(k, n)$ for values of k between 1 and 7.

(h) What would you want to fill into row 0 and column 0 of your table in order to make it consistent with your recurrence. What does this say $P(0, 0)$ should be? We usually define a sum with no terms in it to be zero. Is that consistent with the way the recurrence says we should define $P(0, 0)$? (h)

It is remarkable that there is no known formula for $P(k, n)$, nor is there one for $P(k)$. This section was are devoted to developing methods for computing values of $P(n, k)$ and finding properties of $P(n, k)$ that we can prove even without knowing a formula. Some future sections will attempt to develop other methods.

We have seen that the number of partitions of k into n parts is equal to the number of ways to distribute k identical objects to n recipients so that each receives at least one. If we relax the condition that each recipient receives at least one, then we see that the number of distributions of k identical objects to n recipients is $\sum_{i=1}^{n} P(k, i)$ because if some recipients receive nothing, it does not matter which recipients these are. This completes rows 7 and 8 of our table of distribution problems. The completed table is shown in Table 3.3.4. There are quite a few theorems that you have proved which are summarized by Table 3.3.4. It would be worthwhile to try to write them all down!

3.3. Partitions of Integers

The Twentyfold Way: A Table of Distribution Problems		
k objects and conditions on how they are received	n recipients and mathematical model for distribution	
	Distinct	Identical
1. Distinct no conditions	n^k functions	$\sum_{i=1}^{k} S(n,i)$ set partitions ($\leq n$ parts)
2. Distinct Each gets at most one	$n^{\underline{k}}$ k-element permutations	1 if $k \leq n$; 0 otherwise
3. Distinct Each gets at least one	$S(k,n)n!$ onto functions	$S(k,n)$ set partitions (n parts)
4. Distinct Each gets exactly one	$k! = n!$ permutations	1 if $k = n$; 0 otherwise
5. Distinct, order matters	$(k+n-1)^{\underline{k}}$ ordered functions	$\sum_{i=1}^{n} L(k,i)$ broken permutations ($\leq n$ parts)
6. Distinct, order matters Each gets at least one	$(k)^{\underline{n}}(k-1)^{\underline{k-n}}$ ordered onto functions	$L(k,n) = \binom{k}{n}(k-1)^{\underline{k-n}}$ broken permutations (n parts)
7. Identical no conditions	$\binom{n+k-1}{k}$ multisets	$\sum_{i=1}^{n} P(k,i)$ number partitions ($\leq n$ parts)
8. Identical Each gets at most one	$\binom{n}{k}$ subsets	1 if $k \leq n$; 0 otherwise
9. Identical Each gets at least one	$\binom{k-1}{n-1}$ compositions (n parts)	$P(k,n)$ number partitions (n parts)
10. Identical Each gets exactly one	1 if $k = n$; 0 otherwise	1 if $k = n$; 0 otherwise

Table 3.3.4: The number of ways to distribute k objects to n recipients, with restrictions on how the objects are received

3.3.4 Partitions into distinct parts

Often $Q(k,n)$ is used to denote the number of partitions of k into distinct parts, that is, parts that are different from each other.

Problem 172. Show that
$$Q(k,n) \leq \frac{1}{n!}\binom{k-1}{n-1}.$$

(h)

⇒ **Problem 173.** Show that the number of partitions of 7 into 3 parts equals the number of partitions of 10 into three distinct parts. (h)

⇒ · **Problem 174.** There is a relationship between $P(k, n)$ and $Q(m, n)$ for some other number m. Find the number m that gives you the nicest possible relationship. (h)

· **Problem 175.** Find a recurrence that expresses $Q(k, n)$ as a sum of $Q(k - n, m)$ for appropriate values of m. (h)

⇒ * **Problem 176.** Show that the number of partitions of k into distinct parts equals the number of partitions of k into odd parts. (h)

⇒ * **Problem 177.** Euler showed that if $k \neq \frac{3j^2+j}{2}$, then the number of partitions of k into an even number of distinct parts is the same as the number of partitions of k into an odd number of distinct parts. Prove this, and in the exceptional case find out how the two numbers relate to each other. (h)

3.4 Supplementary Problems

1. Answer each of the following questions with n^k, k^n, $n!$, $k!$, $\binom{n}{k}$, $\binom{k}{n}$, $n^{\underline{k}}$, $k^{\underline{n}}$, $n^{\overline{k}}$, $k^{\overline{n}}$, $\binom{n+k-1}{k}$, $\binom{n+k-1}{n}$, $\binom{n-1}{k-1}$, $\binom{k-1}{n-1}$, or "none of the above".
   (a) In how many ways may we pass out k identical pieces of candy to n children?
   (b) In how many ways may we pass out k distinct pieces of candy to n children?
   (c) In how many ways may we pass out k identical pieces of candy to n children so that each gets at most one? (Assume $k \leq n$.)
   (d) In how many ways may we pass out k distinct pieces of candy to n children so that each gets at most one? (Assume $k \leq n$.)
   (e) In how many ways may we pass out k distinct pieces of candy to n children so that each gets at least one? (Assume $k \geq n$.)
   (f) In how many ways may we pass out k identical pieces of candy to n children so that each gets at least one? (Assume $k \geq n$.)

2. The neighborhood betterment committee has been given r trees to distribute to s families living along one side of a street.

(a) In how many ways can they distribute all of them if the trees are distinct, there are more families than trees, and each family can get at most one?

(b) In how many ways can they distribute all of them if the trees are distinct, any family can get any number, and a family may plant its trees where it chooses?

(c) In how many ways can they distribute all the trees if the trees are identical, there are no more trees than families, and any family receives at most one?

(d) In how many ways can they distribute them if the trees are distinct, there are more trees than families, and each family receives at most one (so there could be some leftover trees)?

(e) In how many ways can they distribute all the trees if they are identical and anyone may receive any number of trees?

(f) In how many ways can all the trees be distributed and planted if the trees are distinct, any family can get any number, and a family must plant its trees in an evenly spaced row along the road?

(g) Answer the question in Part 3.4.2.f assuming that every family must get a tree.

(h) Answer the question in Part 3.4.2.e assuming that each family must get at least one tree.

3. In how many ways can n identical chemistry books, r identical mathematics books, s identical physics books, and t identical astronomy books be arranged on three bookshelves? (Assume there is no limit on the number of books per shelf.)

⇒ 4. One formula for the Lah numbers is

$$L(k, n) = \binom{k}{n}(k-1)^{\underline{k-n}}$$

Find a proof that explains this product.

5. What is the number of partitions of n into two parts?

6. What is the number of partitions of k into $k - 2$ parts?

7. Show that the number of partitions of k into n parts of size at most m equals the number of partitions of $mn - k$ into no more than n parts of size at most $m - 1$.

8. Show that the number of partitions of k into parts of size at most m is equal to the number of partitions of of $k + m$ into m parts.

9. You can say something pretty specific about self-conjugate partitions of k into distinct parts. Figure out what it is and prove it. With that, you should be able to find a relationship between these partitions and partitions whose parts are consecutive integers, starting with 1. What is that relationship?

10. What is $s(k, 1)$?

11. Show that the Stirling numbers of the second kind satisfy the recurrence

$$S(k, n) = \sum_{i=1}^{k} S(k-i, n-1) \binom{n-1}{i-1}.$$

⇒ 12. Let $c(k,n)$ be the number of ways for k children to hold hands to form n circles, where one child clasping his or her hands together and holding them out to form a circle is considered a circle. Find a recurrence for $c(k,n)$. Is the family of numbers $c(k,n)$ related to any of the other families of numbers we have studied? If so, how?

⇒ 13. How many labeled trees on n vertices have exactly four vertices of degree 1?

⇒ 14. The **degree sequence** of a graph is a list of the degrees of the vertices in nonincreasing order. For example the degree sequence of the first graph in Figure 2.3.3 is $(4,3,2,2,1)$. For a graph with vertices labeled 1 through n, the **ordered degree sequence** of the graph is the sequence $d_1, d_2, \ldots, d_n$ in which d_i is the degree of vertex i. For example the ordered degree sequence of the first graph in Figure 2.3.1 is $(1,2,3,3,1,1,2,1)$.

  (a) How many labeled trees are there on n vertices with ordered degree sequence $d_1, d_2, \ldots, d_n$?

  (b) How many labeled trees are there on n vertices with with the degree sequence in which the degree d appears i_d times?

Chapter 4

Generating Functions

4.1 The Idea of Generating Functions

4.1.1 Visualizing Counting with Pictures

Suppose you are going to choose three pieces of fruit from among apples, pears and bananas for a snack. We can symbolically represent all your choices as

🍎🍎🍎 + 🍐🍐🍐 + 🍌🍌🍌 + 🍎🍎🍐 + 🍎🍎🍌 + 🍎🍐🍐 + 🍐🍐🍌 + 🍎🍌🍌 + 🍐🍌🍌 + 🍎🍐🍌.

Here we are using a picture of a piece of fruit to stand for taking a piece of that fruit. Thus 🍎 stands for taking an apple, 🍎🍐 for taking an apple and a pear, and 🍎🍎 for taking two apples. You can think of the plus sign as standing for the "exclusive or," that is, 🍎 + 🍌 would stand for "I take an apple or a banana but not both." To say "I take both an apple and a banana," we would write 🍎🍌. We can extend the analogy to mathematical notation by condensing our statement that we take three pieces of fruit to

$$🍎^3 + 🍐^3 + 🍌^3 + 🍎^2🍐 + 🍎^2🍌 + 🍎🍐^2 + 🍐^2🍌 + 🍎🍌^2 + 🍐🍌^2 + 🍎🍐🍌.$$

In this notation $🍎^3$ stands for taking a multiset of three apples, while $🍎^2🍌$ stands for taking a multiset of two apples and a banana, and so on. What our notation is really doing is giving us a convenient way to list all three element multisets chosen from the set $\{🍎, 🍐, 🍌\}$.[1]

Suppose now that we plan to choose between one and three apples, between one and two pears, and between one and two bananas. In a somewhat clumsy way we could describe our fruit selections as

$$\begin{aligned} &🍎🍐🍌 + 🍎^2🍐🍌 \quad &&+ \cdots + 🍎^2🍐^2🍌 \quad &&+ \cdots + 🍎^2🍐^2🍌^2 \\ &+ 🍎^3🍐🍌 \quad &&+ \cdots + 🍎^3🍐^2🍌 \quad &&+ \cdots + 🍎^3🍐^2🍌^2. \end{aligned} \quad (4.1)$$

---

[1]This approach was inspired by George Pólya's paper "Picture Writing," in the December, 1956 issue of the *American Mathematical Monthly*, page 689. While we are taking a somewhat more formal approach than Pólya, it is still completely in the spirit of his work.

- **Problem 178.** Using an A in place of the picture of an apple, a P in place of the picture of a pear, and a B in place of the picture of a banana, write out the formula similar to Formula (4.1) without any dots for left out terms. (You may use pictures instead of letters if you prefer, but it gets tedious quite quickly!) Now expand the product $(A + A^2 + A^3)(P + P^2)(B + B^2)$ and compare the result with your formula.

- **Problem 179.** Substitute x for all of A, P and B (or for the corresponding pictures) in the formula you got in Problem 178 and expand the result in powers of x. Give an interpretation of the coefficient of x^n.

If we were to expand the formula

$$(\text{🍎} + \text{🍎}^2 + \text{🍎}^3)(\text{🍐} + \text{🍐}^2)(\text{🍌} + \text{🍌}^2). \tag{4.2}$$

we would get Formula (4.1). Thus Formula (4.1) and Formula (4.2) each describe the number of multisets we can choose from the set $\{\text{🍎}, \text{🍐}, \text{🍌}\}$ in which 🍎 appears between 1 and three times and 🍐 and 🍌 each appear once or twice. We interpret Formula (4.1) as describing each individual multiset we can choose, and we interpret Formula (4.2) as saying that we first decide how many apples to take, and then decide how many pears to take, and then decide how many bananas to take. At this stage it might seem a bit magical that doing ordinary algebra with the second formula yields the first, but in fact we could define addition and multiplication with these pictures more formally so we could explain in detail why things work out. However since the pictures are for motivation, and are actually difficult to write out on paper, it doesn't make much sense to work out these details. We will see an explanation in another context later on.

4.1.2 Picture functions

As you've seen, in our descriptions of ways of choosing fruits, we've treated the pictures of the fruit as if they are variables. You've also likely noticed that it is much easier to do algebraic manipulations with letters rather than pictures, simply because it is time consuming to draw the same picture over and over again, while we are used to writing letters quickly. In the theory of generating functions, we associate variables or polynomials or even power series with members of a set. There is no standard language describing how we associate variables with members of a set, so we shall invent[2] some. By a **picture** of a member of a set we will mean a variable, or perhaps a product of powers of variables (or even a sum of products of powers of variables). A function that assigns a picture $P(s)$ to each member s of a set S will be called a **picture function**. The **picture enumerator** for a picture function P defined on a set S will be

$$E_P(S) = \sum_{s: s \in S} P(s).$$

---

[2] We are really adapting language introduced by George Pólya.

We choose this language because the picture enumerator lists, or enumerates, all the elements of S according to their pictures. Thus Formula (4.1) is the picture enumerator the set of all multisets of fruit with between one and three apples, one and two pears, and one and two bananas.

- **Problem 180.** How would you write down a polynomial in the variable A that says you should take between zero and three apples?

- **Problem 181.** How would you write down a picture enumerator that says we take between zero and three apples, between zero and three pears, and between zero and three bananas?

- **Problem 182.** (Used in Chapter 6.) Notice that when we used A^2 to stand for taking two apples, and P^3 to stand for taking three pears, then we used the product A^2P^3 to stand for taking two apples and three pears. Thus we have chosen the picture of the ordered pair (2 apples, 3 pears) to be the product of the pictures of a multiset of two apples and a multiset of three pears. Show that if S_1 and S_2 are sets with picture functions P_1 and P_2 defined on them, and if we define the picture of an ordered pair $(x_1, x_2) \in S_1 \times S_2$ to be $P((x_1, x_2)) = P_1(x_1)P_2(x_2)$, then the picture enumerator of P on the set $S_1 \times S_2$ is $E_{P_1}(S_1)E_{P_2}(S_2)$. We call this the **product principle for picture enumerators**.

4.1.3 Generating functions

- **Problem 183.** Suppose you are going to choose a snack of between zero and three apples, between zero and three pears, and between zero and three bananas. Write down a polynomial in one variable x such that the coefficient of x^n is the number of ways to choose a snack with n pieces of fruit. (h)

- **Problem 184.** Suppose an apple costs 20 cents, a banana costs 25 cents, and a pear costs 30 cents. What should you substitute for A, P, and B in Problem 181 in order to get a polynomial in which the coefficient of x^n is the number of ways to choose a selection of fruit that costs n cents? (h)

- **Problem 185.** Suppose an apple has 40 calories, a pear has 60 calories, and a banana has 80 calories. What should you substitute for A, P, and B in Problem 181 in order to get a polynomial in which the coefficient of x^n is the number of ways to choose a selection of fruit with a total of n calories?

- **Problem 186.** We are going to choose a subset of the set $[n] = \{1, 2, \ldots, n\}$. Suppose we use x_1 to be the picture of choosing 1 to be in our subset. What is the picture enumerator for either choosing 1 or not choosing 1? Suppose that for each i between 1 and n, we use x_i to be the picture of choosing i to be in our subset. What is the picture enumerator for either choosing i or not choosing i to be in our subset? What is the picture enumerator for all possible choices of subsets of $[n]$? What should we substitute for x_i in order to get a polynomial in x such that the coefficient of x^k is the number of ways to choose a k-element subset of n? What theorem have we just reproved (a special case of)? (h)

In Problem 186 we see that we can think of the process of expanding the polynomial $(1+x)^n$ as a way of "generating" the binomial coefficients $\binom{n}{k}$ as the coefficients of x^k in the expansion of $(1+x)^n$. For this reason, we say that $(1+x)^n$ is the "generating function" for the binomial coefficients $\binom{n}{k}$. More generally, the **generating function** for a sequence a_i, defined for i with $0 \leq i \leq n$ is the expression $\sum_{i=0}^{n} a_i x^i$, and the **generating function** for the sequence a_i with $i \geq 0$ is the expression $\sum_{i=0}^{\infty} a_i x^i$. This last expression is an example of a power series. In calculus it is important to think about whether a power series converges in order to determine whether or not it represents a function. In a nice twist of language, even though we use the phrase generating function as the name of a power series in combinatorics, we don't require the power series to actually represent a function in the usual sense, and so we don't have to worry about convergence.[3] Instead we think of a power series as a convenient way of representing the terms of a sequence of numbers of interest to us. The only justification for saying that such a representation is convenient is because of the way algebraic properties of power series capture some of the important properties of some sequences that are of combinatorial importance. The remainder of this chapter is devoted to giving examples of how the algebra of power series reflects combinatorial ideas.

Because we choose to think of power series as strings of symbols that we manipulate by using the ordinary rules of algebra and we choose to ignore issues of convergence, we have to avoid manipulating power series in a way that would require us to add infinitely many real numbers. For example, we cannot make the substitution of $y + 1$ for x in the power series $\sum_{i=0}^{\infty} x^i$, because in order to interpret $\sum_{i=0}^{\infty} (y+1)^i$ as a power series we would have to apply the binomial theorem to each of the $(y+1)^i$ terms, and then collect like terms, giving us infinitely many

---

[3]In the evolution of our current mathematical terminology, the word function evolved through several meanings, starting with very imprecise meanings and ending with our current rather precise meaning. The terminology "generating function" may be thought of as an example of one of the earlier usages of the term function.

ones added together as the coefficient of y^0, and in fact infinitely many numbers added together for the coefficient of any y^i. (On the other hand, it would be fine to substitute $y + y^2$ for x. Can you see why?)

4.1.4 Power series

For now, most of our uses of power series will involve just simple algebra. Since we use power series in a different way in combinatorics than we do in calculus, we should review a bit of the algebra of power series.

- **Problem 187.** In the polynomial $(a_0 + a_1x + a_2x^2)(b_0 + b_1x + b_2x^2 + b_3x^3)$, what is the coefficient of x^2? What is the coefficient of x^4?

- **Problem 188.** In Problem 187 why is there a b_0 and a b_1 in your expression for the coefficient of x^2 but there is not a b_0 or a b_1 in your expression for the coefficient of x^4? What is the coefficient of x^4 in

  $$(a_0 + a_1x + a_2x^2 + a_3x^3 + a_4x^4)(b_0 + b_1x + b_2x^2 + b_3x^3 + b_4x^4)?$$

  Express this coefficient in the form

  $$\sum_{i=0}^{4} \text{something,}$$

  where the something is an expression you need to figure out. Now suppose that $a_3 = 0$, $a_4 = 0$ and $b_4 = 0$. To what is your expression equal after you substitute these values? In particular, what does this have to do with Problem 187? (h)

- **Problem 189.** The point of the Problems 187 and Problem 188 is that so long as we are willing to assume $a_i = 0$ for $i > n$ and $b_j = 0$ for $j > m$, then there is a very nice formula for the coefficient of x^k in the product

  $$\left(\sum_{i=0}^{n} a_i x^i\right)\left(\sum_{j=0}^{m} b_j x^j\right).$$

  Write down this formula explicitly. (h)

- **Problem 190.** Assuming that the rules you use to do arithmetic with polynomials apply to power series, write down a formula for the coefficient of

x^k in the product

$$\left(\sum_{i=0}^{\infty} a_i x^i\right)\left(\sum_{j=0}^{\infty} b_j x^j\right).$$

(h)

We use the expression you obtained in Problem 190 to *define* the product of power series. That is, we define the product

$$\left(\sum_{i=0}^{\infty} a_i x^i\right)\left(\sum_{j=0}^{\infty} b_j x^j\right)$$

to be the power series $\sum_{k=0}^{\infty} c_k x^k$, where c_k is the expression you found in Problem 190. Since you derived this expression by using the usual rules of algebra for polynomials, it should not be surprising that the product of power series satisfies these rules.[4]

4.1.5 Product principle for generating functions

Each time that we converted a picture function to a generating function by substituting x or some power of x for each picture, the coefficient of x had a meaning that was significant to us. For example, with the picture enumerator for selecting between zero and three each of apples, pears, and bananas, when we substituted x for each of our pictures, the exponent i in the power x^i is the number of pieces of fruit in the fruit selection that led us to x^i. After we simplify our product by collecting together all like powers of x, the coefficient of x^i is the number of fruit selections that use i pieces of fruit. In the same way, if we substitute x^c for a picture, where c is the number of calories in that particular kind of fruit, then the i in an x^i term in our generating function stands for the number of calories in a fruit selection that gave rise to x^i, and the coefficient of x^i in our generating function is the number of fruit selections with i calories. The product principle of picture enumerators translates directly into a product principle for generating functions.

- **Problem 191.** Suppose that we have two sets S_1 and S_2. Let v_1 (v stands for value) be a function from S_1 to the nonnegative integers and let v_2 be a function from S_2 to the nonnegative integers. Define a new function v on the set $S_1 \times S_2$ by $v(x_1, x_2) = v_1(x_1) + v_2(x_2)$. Suppose further that $\sum_{i=0}^{\infty} a_i x^i$ is the generating function for the number of elements x_1 of S_1 of value i, that is with $v_1(x_1) = i$. Suppose also that $\sum_{j=0}^{\infty} b_j x^j$ is the generating function for the number of elements x_2 of S_2 of value j, that is with $v_2(x_2) = j$. Prove

---

[4]Technically we should explicitly state these rules and prove that they are all valid for power series multiplication, but it seems like overkill at this point to do so!

that the coefficient of x^k in

$$\left(\sum_{i=0}^{\infty} a_i x^i\right)\left(\sum_{j=0}^{\infty} b_j x^j\right)$$

is the number of ordered pairs (x_1, x_2) in $S_1 \times S_2$ with total value k, that is with $v_1(x_1) + v_2(x_2) = k$. This is called the **product principle for generating functions**. (h)

Problem 191 may be extended by mathematical induction to prove our next theorem.

Theorem 4.1.1. *If $S_1, S_2, \ldots, S_n$ are sets with a value function v_i from S_i to the nonnegative integers for each i and $f_i(x)$ is the generating function for the number of elements of S_i of each possible value, then the generating function for the number of n-tuples of each possible value is $\prod_{i=1}^{n} f_i(x)$.*

4.1.6 The extended binomial theorem and multisets

- **Problem 192.** Suppose once again that i is an integer between 1 and n.

    (a) What is the generating function in which the coefficient of x^k is 1? This series is an example of what is called an **infinite geometric series**. In the next part of this problem, it will be useful to interpret the coefficient one as the number of multisets of size k chosen from the singleton set $\{i\}$. Namely, there is only one way to choose a multiset of size k from $\{i\}$: choose i exactly k times.

    (b) Express the generating function in which the coefficient of x^k is the number of multisets chosen from $[n]$ as a power of a power series. What does Problem 125 (in which your answer could be expressed as a binomial coefficient) tell you about what this generating function equals? (h)

○ **Problem 193.** What is the product $(1-x) \sum_{k=0}^{n} x^k$? What is the product

$$(1-x) \sum_{k=0}^{\infty} x^k ?$$

Problem 194. Express the generating function for the number of multisets of size k chosen from $[n]$ (where n is fixed but k can be any nonnegative integer) as a 1 over something relatively simple.

- **Problem 195.** Find a formula for $(1+x)^{-n}$ as a power series whose coefficients involve binomial coefficients. What does this formula tell you about how we should define $\binom{-n}{k}$ when n is positive? (h)

Problem 196. If you define $\binom{-n}{k}$ in the way you described in Problem 195, you can write down a version of the binomial theorem for $(x+y)^n$ that is valid for both nonnegative and negative values of n. Do so. This is called the **extended binomial theorem**. Write down a special case with n negative, like $n = -3$, to see an interesting surprise that suggests why we do not use this formula later on.

Problem 197. Write down the generating function for the number of ways to distribute identical pieces of candy to three children so that no child gets more than 4 pieces. Write this generating function as a quotient of polynomials. Using both the extended binomial theorem and the original binomial theorem, find out in how many ways we can pass out exactly ten pieces. (h)

- **Problem 198.** What is the generating function for the number of multisets chosen from an n-element set so that each element appears at least j times and less than m times? Write this generating function as a quotient of polynomials, then as a product of a polynomial and a power series. (h)

⇒ **Problem 199.** Recall that a tree is determined by its edge set. Suppose you have a tree on n vertices, say with vertex set $[n]$. We can use x_i as the picture of vertex i and $x_i x_j$ as the picture of the edge $x_i x_j$. Then one possible picture of the tree T is the product $P(T) = \prod_{\{i,j\}: i \text{ and } j \text{ are adjacent}} x_i x_j$.

   (a) Explain why the picture of a tree is also $\prod_{i=1}^{n} x_i^{\deg(i)}$.

   (b) Write down the picture enumerators for trees on two, three, and four vertices. Factor them as completely as possible.

   (c) Explain why $x_1 x_2 \cdots x_n$ is a factor of the picture of a tree on n vertices.

   (d) Write down the picture of a tree on five vertices with one vertex of degree four, say vertex i. If a tree on five vertices has a vertex of degree three, what are the possible degrees of the other vertices. What can you say about the picture of a tree with a vertex of degree three? If a tree on five vertices has no vertices of degree three or four, how many vertices of degree two does it have? What can you say about its picture? Write down the picture enumerator for trees on five vertices.

(e) Find a (relatively) simple polynomial expression for the picture enumerator $\sum_{T:\ T \text{ is a tree on } [n]} P(T)$. Prove it is correct. (h)

(f) The enumerator for trees by degree sequence is the sum over all trees of $x^{d_1} x^{d_2} \cdots x^{d_n}$, where d_i is the degree of vertex i. What is the enumerator by degree sequence for trees on the vertex set $[n]$?

4.2 Generating functions for integer partitions

- **Problem 200.** If we have five identical pennies, five identical nickels, five identical dimes, and five identical quarters, give the picture enumerator for the combinations of coins we can form and convert it to a generating function for the number of ways to make k cents with the coins we have. Do the same thing assuming we have an unlimited supply of pennies, nickels, dimes, and quarters. (h)

- **Problem 201.** Recall that a partition of an integer k is a multiset of numbers that adds to k. In Problem 200 we found the generating function for the number of partitions of an integer into parts of size 1, 5, 10, and 25. When working with generating functions for partitions, it is becoming standard to use q rather than x as the variable in the generating function. Write your answers in this notation.[a]

  (a) Give the generating function for the number partitions of an integer into parts of size one through ten. (h)

  (b) Give the generating function for the number of partitions of an integer k into parts of size at most m, where m is fixed but k may vary. Notice this is the generating function for partitions whose Young diagram fits into the space between the line $x = 0$ and the line $x = m$ in a coordinate plane. (We assume the boxes in the Young diagram are one unit by one unit.) (h)

  ---
  [a]The reason for this change in the notation relates to the subject of finite fields in abstract algebra, where q is the standard notation for the size of a finite field. While we will make no use of this connection, it will be easier for you to read more advanced work if you get used to the different notation.

- **Problem 202.** In Problem 201.b you gave the generating function for the number of partitions of an integer into parts of size at most m. Explain why this is also the generating function for partitions of an integer into at most m parts. Notice that this is the generating function for the number of partitions whose Young diagram fits into the space between the line $y = 0$ and the line $y = m$. (h)

- **Problem 203.** When studying partitions of integers, it is inconvenient to restrict ourselves to partitions with at most m parts or partitions with maximum part size m.

  (a) Give the generating function for the number of partitions of an integer into parts of any size. Don't forget to use q rather than x as your variable. (h)

  (b) Find the coefficient of q^4 in this generating function. (h)

  (c) find the coefficient of q^5 in this generating function.

  (d) This generating function involves an infinite product. Describe the process you would use to expand this product into as many terms of a power series as you choose. (h)

  (e) Rewrite any power series that appear in your product as quotients of polynomials or as integers divided by polynomials.

$\Rightarrow$ **Problem 204.** In Problem 203, we multiplied together infinitely many power series. Here are two notations for infinite products that look rather similar:

$$\prod_{i=1}^{\infty} 1 + x + x^2 + \cdots + x^i \quad \text{and} \quad \prod_{i=1}^{\infty} 1 + x^i + x^{2i} + \cdots + x^{i^2}.$$

However, one makes sense and one doesn't. Figure out which one makes sense and explain why it makes sense and the other one doesn't. If we want a product of the form

$$\prod_{i=1}^{\infty} 1 + p_i(x),$$

where each $p_i(x)$ is a nonzero polynomial in x to make sense, describe a relatively simple assumption about the polynomials $p_i(x)$ that will make the product make sense. If we assumed the terms $p_i(x)$ were nonzero power series, is there a relatively simple assumption we could make about them in order to make the product make sense? (Describe such a condition or explain why you think there couldn't be one.) (h)

- **Problem 205.** What is the generating function (using q for the variable) for the number of partitions of an integer in which each part is even? (h)

- **Problem 206.** What is the generating function (using q as the variable) for the number of partitions of an integer into distinct parts, that is, in which each part is used at most once? (h)

- **Problem 207.** Use generating functions to explain why the number of partitions of an integer in which each part is used an even number of times equals the generating function for the number of partitions of an integer in which each part is even. (h)

Problem 208. Use the fact that

$$\frac{1-q^{2i}}{1-q^i} = 1 + q^i$$

and the generating function for the number of partitions of an integer into distinct parts to show how the number of partitions of an integer k into distinct parts is related to the number of partitions of an integer k into odd parts. (h)

Problem 209. Write down the generating function for the number of ways to partition an integer into parts of size no more than m, each used an odd number of times. Write down the generating function for the number of partitions of an integer into parts of size no more than m, each used an even number of times. Use these two generating functions to get a relationship between the two sequences for which you wrote down the generating functions. (h)

⇒ **Problem 210.** In Problem 201.b and Problem 202 you gave the generating functions for, respectively, the number of partitions of k into parts the largest of which is at most m and for the number of partitions of k into at most m parts. In this problem we will give the generating function for the number of partitions of k into at most n parts, the largest of which is at most m. That is we will analyze $\sum_{i=0}^{\infty} a_k q^k$ where a_k is the number of partitions of k into at most n parts, the largest of which is at most m. Geometrically, it is the generating function for partitions whose Young diagram fits into an m by n rectangle, as in Problem 167. This generating function has significant analogs to the binomial coefficient $\binom{m+n}{n}$, and so it is denoted by $\begin{bmatrix} m+n \\ n \end{bmatrix}_q$. It is called a q-**binomial coefficient**.

(a) Compute $\begin{bmatrix} 4 \\ 2 \end{bmatrix}_q = \begin{bmatrix} 2+2 \\ 2 \end{bmatrix}_q$. (h)

(b) Find explicit formulas for $\begin{bmatrix} n \\ 1 \end{bmatrix}_q$ and $\begin{bmatrix} n \\ n-1 \end{bmatrix}_q$. (h)

(c) How are $\begin{bmatrix} m+n \\ n \end{bmatrix}_q$ and $\begin{bmatrix} m+n \\ n \end{bmatrix}_q$ related? Prove it. (Note this is the same as asking how $\begin{bmatrix} r \\ s \end{bmatrix}_q$ and $\begin{bmatrix} r \\ r-s \end{bmatrix}_q$ are related.) (h)

(d) So far the analogy to $\binom{m+n}{n}$ is rather thin! If we had a recurrence like the Pascal recurrence, that would demonstrate a real analogy. Is $\begin{bmatrix} m+n \\ n \end{bmatrix}_q = \begin{bmatrix} m+n-1 \\ n-1 \end{bmatrix}_q + \begin{bmatrix} m+n-1 \\ n \end{bmatrix}_q$?

(e) Recall the two operations we studied in Problem 171.

   (i) The largest part of a partition counted by $\begin{bmatrix} m+n \\ n \end{bmatrix}_q$ is either m or is less than or equal to $m - 1$. In the second case, the partition fits into a rectangle that is at most $m - 1$ units wide and at most n units deep. What is the generating function for partitions of this type? In the first case, what kind of rectangle does the partition we get by removing the largest part sit in? What is the generating function for partitions that sit in this kind of rectangle? What is the generating function for partitions that sit in this kind of rectangle after we remove a largest part of size m? What recurrence relation does this give you?

   (ii) What recurrence do you get from the other operation we studied in Problem 171?

   (iii) It is quite likely that the two recurrences you got are different. One would expect that they might give different values for $\begin{bmatrix} m+n \\ n \end{bmatrix}_q$. Can you resolve this potential conflict? (h)

(f) Define $[n]_q$ to be $1 + q + \cdots + q^{n-1}$ for $n > 0$ and $[0]_q = 1$. We read this simply as n-sub-q. Define $[n]!_q$ to be $[n]_q[n-1]_q \cdots [3]_q[2]_q[1]_q$. We read this as n cue-torial, and refer to it as a q-**ary factorial**. Show that

$$\begin{bmatrix} m+n \\ n \end{bmatrix}_q = \frac{[m+n]!_q}{[m]!_q[n]!_q}.$$

(h)

(g) Now think of q as a variable that we will let approach 1. Find an explicit formula for

   (i) $\lim_{q \to 1} [n]_q$.

(ii) $\lim_{q \to 1} [n]!_q$.

(iii) $\lim_{q \to 1} \begin{bmatrix} m+n \\ n \end{bmatrix}_q$.

Why is the limit in Part iii equal to the number of partitions (of any number) with at most n parts all of size most m? Can you explain bijectively why this quantity equals the formula you got? (h)

* **(h)** What happens to $\begin{bmatrix} m+n \\ n \end{bmatrix}_q$ if we let q approach -1? (h)

4.3 Generating Functions and Recurrence Relations

Recall that a recurrence relation for a sequence a_n expresses a_n in terms of values a_i for $i < n$. For example, the equation $a_i = 3a_{i-1} + 2^i$ is a first order linear constant coefficient recurrence.

4.3.1 How generating functions are relevant

Algebraic manipulations with generating functions can sometimes reveal the solutions to a recurrence relation.

- **Problem 211.** Suppose that $a_i = 3a_{i-1} + 3^i$.

    (a) Multiply both sides by x^i and sum both the left hand side and right hand side from $i = 1$ to infinity. In the left-hand side use the fact that

    $$\sum_{i=1}^{\infty} a_i x^i = (\sum_{i=0}^{\infty} x^i) - a_0$$

    and in the right hand side, use the fact that

    $$\sum_{i=1}^{\infty} a_{i-1} x^i = x \sum_{i=1}^{\infty} a_i x^{i-1} = x \sum_{j=0}^{\infty} a_j x^j = x \sum_{i=0}^{\infty} a_i x^i$$

    (where we substituted j for $i-1$ to see explicitly how to change the limits of summation, a surprisingly useful trick) to rewrite the equation in terms of the power series $\sum_{i=0}^{\infty} a_i x^i$. Solve the resulting equation for the power series $\sum_{i=0}^{\infty} a_i x^i$. You can save a lot of writing by using a variable like y to stand for the power series.

    (b) Use the previous part to get a formula for a_i in terms of a_0.

    (c) Now suppose that $a_i = 3a_{i-1} + 2^i$. Repeat the previous two steps for this recurrence relation. (There is a way to do this part using what you already know. Later on we shall introduce yet another way to deal with the kind of generating function that arises here.) (h)

○ **Problem 212.** Suppose we deposit $5000 in a savings certificate that pays ten percent interest and also participate in a program to add $1000 to the certificate at the end of each year (from the end of the first year on) that follows (also subject to interest.) Assuming we make the $5000 deposit at the end of year 0, and letting a_i be the amount of money in the account at the end of year i, write a recurrence for the amount of money the certificate is worth at the end of year n. Solve this recurrence. How much money do we have in the account (after our year-end deposit) at the end of ten years? At the end of 20 years?

4.3.2 Fibonacci numbers

The sequence of problems that follows (culminating in Problem 222) describes a number of hypotheses we might make about a fictional population of rabbits. We use the example of a rabbit population for historic reasons; our goal is a classical sequence of numbers called Fibonacci numbers. When Fibonacci[5] introduced them, he did so with a fictional population of rabbits.

4.3.3 Second order linear recurrence relations

- **Problem 213.** Suppose we start (at the end of month 0) with 10 pairs of baby rabbits, and that after baby rabbits mature for one month they begin to reproduce, with each pair producing two new pairs at the end of each month afterwards. Suppose further that over the time we observe the rabbits, none die. Let a_n be the number of rabbits we have at the end of month n. Show that $a_n = a_{n-1} + 2a_{n-2}$. This is an example of a **second order** *linear* recurrence with constant coefficients. Using a method similar to that of Problem 211, show that
$$\sum_{i=0}^{\infty} a_i x^i = \frac{10}{1-x-2x^2}.$$
This gives us the generating function for the sequence a_i giving the population in month i; shortly we shall see a method for converting this to a solution to the recurrence.

- **Problem 214.** In Fibonacci's original problem, each pair of mature rabbits produces one new pair at the end of each month, but otherwise the situation is the same as in Problem 213. Assuming that we start with one pair of baby rabbits (at the end of month 0), find the generating function for the number of pairs of rabbits we have at the end on n months. (h)

---

[5]Apparently Leanardo de Pisa was given the name Fibonacci posthumously. It is a shortening of "son of Bonacci" in Italian.

⇒ **Problem 215.** Find the generating function for the solutions to the recurrence
$$a_i = 5a_{i-1} - 6a_{i-2} + 2^i.$$

The recurrence relations we have seen in this section are called **second order** because they specify a_i in terms of a_{i-1} and a_{i-2}, they are called **linear** because a_{i-1} and a_{i-2} each appear to the first power, and they are called **constant coefficient recurrences** because the coefficients in front of a_{i-1} and a_{i-2} are constants.

4.3.4 Partial fractions

The generating functions you found in the previous section all can be expressed in terms of the reciprocal of a quadratic polynomial. However without a power series representation, the generating function doesn't tell us what the sequence is. It turns out that whenever you can factor a polynomial into linear factors (and over the complex numbers such a factorization always exists) you can use that factorization to express the reciprocal in terms of power series.

- **Problem 216.** Express $\frac{1}{x-3} + \frac{2}{x-2}$ as a single fraction.

○ **Problem 217.** In Problem 216 you see that when we added numerical multiples of the reciprocals of first degree polynomials we got a fraction in which the denominator is a quadratic polynomial. This will always happen unless the two denominators are multiples of each other, because their least common multiple will simply be their product, a quadratic polynomial. This leads us to ask whether a fraction whose denominator is a quadratic polynomial can always be expressed as a sum of fractions whose denominators are first degree polynomials. Find numbers c and d so that
$$\frac{5x+1}{(x-3)(x+5)} = \frac{c}{x-3} + \frac{d}{x+5}.$$
(h)

- **Problem 218.** In Problem 217 you may have simply guessed at values of c and d, or you may have solved a system of equations in the two unknowns c and d. Given constants a, b, r_1, and r_2 (with $r_1 \neq r_2$), write down a system of equations we can solve for c and d to write
$$\frac{ax+b}{(x-r_1)(x-r_2)} = \frac{c}{x-r_1} + \frac{d}{x-r_2}.$$
(h)

Writing down the equations in Problem 218 and solving them is called the **method of partial fractions**. This method will let you find power series expansions for generating functions of the type you found in Problems 213 to Problem 215. However you have to be able to factor the quadratic polynomials that are in the denominators of your generating functions.

- **Problem 219.** Use the method of partial fractions to convert the generating function of Problem 213 into the form
$$\frac{c}{x-r_1} + \frac{d}{x-r_2}.$$
Use this to find a formula for a_n.

- **Problem 220.** Use the quadratic formula to find the solutions to $x^2 + x - 1 = 0$, and use that information to factor $x^2 + x - 1$.

- **Problem 221.** Use the factors you found in Problem 220 to write
$$\frac{1}{x^2 + x - 1}$$
in the form
$$\frac{c}{x-r_1} + \frac{d}{x-r_2}.$$
(h)

- **Problem 222.**

    (a) Use the partial fractions decomposition you found in Problem 220 to write the generating function you found in Problem 214 in the form
    $$\sum_{n=0}^{\infty} a_n x^i$$
    and use this to give an explicit formula for a_n. (h)

    (b) When we have $a_0 = 1$ and $a_1 = 1$, i.e. when we start with one pair of baby rabbits, the numbers a_n are called **Fibonacci Numbers**. Use either the recurrence or your final formula to find a_2 through a_8. Are you amazed that your general formula produces integers, or for that matter produces rational numbers? Why does the recurrence equation tell you that the Fibonacci numbers are all integers?

(c) Explain why there is a real number b such that, for large values of n, the value of the nth Fibonacci number is almost exactly (but not quite) some constant times b^n. (Find b and the constant.)

(d) Find an algebraic explanation (not using the recurrence equation) of what happens to make the square roots of five go away. (h)

(e) As a challenge (which the author has not yet done), see if you can find a way to show algebraically (not using the recurrence relation, but rather the formula you get by removing the square roots of five) that the formula for the Fibonacci numbers yields integers.

Problem 223. Solve the recurrence $a_n = 4a_{n-1} - 4a_{n-2}$.

4.3.5 Catalan Numbers

⇒ **Problem 224.**

(a) Using either lattice paths or diagonal lattice paths, explain why the Catalan Number c_n satisfies the recurrence

$$c_n = \sum_{i=1}^{n-1} c_{i-1} c_{n-i}.$$

(h)

(b) Show that if we use y to stand for the power series $\sum_{n=0}^{\infty} c_n x^n$, then we can find y by solving a quadratic equation. Find y. (h)

(c) Taylor's theorem from calculus tells us that the extended binomial theorem

$$(1+x)^r = \sum_{i=0}^{\infty} \binom{r}{i} x^i$$

holds for any number real number r, where $\binom{r}{i}$ is defined to be

$$\frac{r^{\underline{i}}}{i!} = \frac{r(r-1)\cdots(r-i+1)}{i!}.$$

Use this and your solution for y (note that of the two possible values for y that you get from the quadratic formula, only one gives an actual power series) to get a formula for the Catalan numbers. (h)

4.4 Supplementary Problems

⇒ * **1.** What is the generating function for the number of ways to pass out k pieces of candy from an unlimited supply of identical candy to n children (where n is fixed) so that each child gets between three and six pieces of candy (inclusive)? Use the fact that
$$(1 + x + x^2 + x^3)(1 - x) = 1 - x^4$$
to find a formula for the number of ways to pass out the candy.

○ **2.**

(a) In paying off a mortgage loan with initial amount A, annual interest rate $p\%$ on a monthly basis with a monthly payment of m, what recurrence describes the amount owed after n months of payments in terms of the amount owed after $n - 1$ months? Some technical details: You make the first payment after one month. The amount of interest included in your monthly payment is $.01p/12$. This interest rate is applied to the amount you owed immediately after making your last monthly payment.

(b) Find a formula for the amount owed after n months.

(c) Find a formula for the number of months needed to bring the amount owed to zero. Another technical point: If you were to make the standard monthly payment m in the last month, you might actually end up owing a negative amount of money. Therefore it is ok if the result of your formula for the number of months needed gives a non-integer number of months. The bank would just round up to the next integer and adjust your payment so your balance comes out to zero.

(d) What should the monthly payment be to pay off the loan over a period of 30 years?

⇒ **3.** We have said that for nonnegative i and positive n we want to define $\binom{-n}{i}$ to be $\binom{n+i-1}{i}$. If we want the Pascal recurrence to be valid, how should we define $\binom{-n}{-i}$ when n and i are both positive?

⇒ **4.** Find a recurrence relation for the number of ways to divide a convex n-gon into triangles by means of non-intersecting diagonals. How do these numbers relate to the Catalan numbers?

⇒ **5.** How does $\sum_{k=0}^{n} \binom{n-k}{k}$ relate to the Fibonacci Numbers?

6. Let m and n be fixed. Express the generating function for the number of k-element multisets of an n-element set such that no element appears more than m times as a quotient of two polynomials. Use this expression to get a formula for the number of k-element multisets of an n-element set such that no element appears more than m times.

7. One natural but oversimplified model for the growth of a tree is that all new wood grows from the previous year's growth and is proportional to it in amount. To be more precise, assume that the (total) length of the new growth in a given year is the constant c times the (total) length of new growth in the previous year.

Write down a recurrence for the total length a_n of all the branches of the tree at the end of growing season n. Find the general solution to your recurrence relation. Assume that we begin with a one meter cutting of new wood (from the previous year) which branches out and grows a total of two meters of new wood in the first year. What will the total length of all the branches of the tree be at the end of n years?

$\Rightarrow$ 8. (Relevant to Appendix C) We have some chairs which we are going to paint with red, white, blue, green, yellow and purple paint. Suppose that we may paint any number of chairs red or white, that we may paint at most one chair blue, at most three chairs green, only an even number of chairs yellow, and only a multiple of four chairs purple. In how many ways may we paint n chairs?

9. What is the generating function for the number of partitions of an integer in which each part is used at most m times? Why is this also the generating function for partitions in which consecutive parts (in a decreasing list representation) differ by at most m and the smallest part is also at most m?

Chapter 5

The Principle of Inclusion and Exclusion

5.1 The size of a union of sets

One of our very first counting principles was the **sum principle** which says that the size of a union of disjoint sets is the sum of their sizes. Computing the size of overlapping sets requires, quite naturally, information about how they overlap. Taking such information into account will allow us to develop a powerful extension of the sum principle known as the "principle of inclusion and exclusion."

5.1.1 Unions of two or three sets

○ **Problem 225.** In a biology lab study of the effects of basic fertilizer ingredients on plants, 16 plants are treated with potash, 16 plants are treated with phosphate, and among these plants, eight are treated with both phosphate and potash. No other treatments are used. How many plants receive at least one treatment? If 32 plants are studied, how many receive no treatment?

+ **Problem 226.** Give a formula for the size of the union $A \cup B$ of two sets A in terms of the sizes $|A|$ of A, $|B|$ of B, and $|A \cap B|$ of $A \cap B$. If A and B are subsets of some "universal" set U, express the size of the complement $U - (A \cup B)$ in terms of the sizes $|U|$ of U, $|A|$ of A, $|B|$ of B, and $|A \cap B|$ of $A \cap B$. (h)

○ **Problem 227.** In Problem 225, there were just two fertilizers used to treat the sample plants. Now suppose there are three fertilizer treatments, and 15 plants are treated with nitrates, 16 with potash, 16 with phosphate, 7 with nitrate and potash, 9 with nitrate and phosphate, 8 with potash and phosphate and 4 with all three. Now how many plants have been treated? If 32 plants were studied, how many received no treatment at all?

- **Problem 228.** Give a formula for the size of $A \cup B \cup C$ in terms of the sizes of A, B, C and the intersections of these sets. (h)

5.1.2 Unions of an arbitrary number of sets

- **Problem 229.** Conjecture a formula for the size of a union of sets

$$A_1 \cup A_2 \cup \cdots \cup A_n = \bigcup_{i=1}^{n} A_i$$

in terms of the sizes of the sets A_i and their intersections.

The difficulty of generalizing Problem 228 to Problem 229 is not likely to be one of being able to see what the right conjecture is, but of finding a good notation to express your conjecture. In fact, it would be easier for some people to express the conjecture in words than to express it in a notation. Here is some notation that will make your task easier. Let us define

$$\bigcap_{i:i \in I} A_i$$

to mean the intersection over all elements i in the set I of A_i. Thus

$$\bigcap_{i:i \in \{1,3,4,6\}} = A_1 \cap A_3 \cap A_4 \cap A_6. \tag{5.1}$$

This kind of notation, consisting of an operator with a description underneath of the values of a dummy variable of interest to us, can be extended in many ways. For example

$$\sum_{I: I \subseteq \{1,2,3,4\},\ |I|=2} |\cap_{i \in I} A_i| = |A_1 \cap A_2| + |A_1 \cap A_3| + |A_1 \cap A_4|$$

$$+ |A_2 \cap A_3| + |A_2 \cap A_4| + |A_3 \cap A_4|. \tag{5.2}$$

- **Problem 230.** Use notation something like that of Equation (5.1) and Equation (5.2) to express the answer to Problem 229. Note there are many different correct ways to do this problem. Try to write down more than one and choose the nicest one you can. Say why you chose it (because your view of what makes a formula nice may be different from somebody else's). The nicest formula won't necessarily involve all the elements of Equations (5.1) and (5.2).

- **Problem 231.** A group of n students goes to a restaurant carrying backpacks. The manager invites everyone to check their backpack at the check desk and everyone does. While they are eating, a child playing in the check room randomly moves around the claim check stubs on the backpacks. We will try to compute the probability that, at the end of the meal, at least one student receives his or her own backpack. This probability is the fraction of the total number of ways to return the backpacks in which at least one student gets his or her own backpack back.

    (a) What is the total number of ways to pass back the backpacks?

    (b) In how many of the distributions of backpacks to students does at least one student get his or her own backpack? (h)

    (c) What is the probability that at least one student gets the correct backpack?

    (d) What is the probability that no student gets his or her own backpack?

    ⇒ (e) As the number of students becomes large, what does the probability that no student gets the correct backpack approach?

Problem 231 is "classically" called the **hatcheck problem**; the name comes from substituting hats for backpacks. If is also sometimes called the **derangement problem**. A **derangement** of an n-element set is a permutation of that set (thought of as a bijection) that maps no element of the set to itself. One can think of a way of handing back the backpacks as a permutation f of the students: $f(i)$ is the owner of the backpack that student i receives. Then a derangement is a way to pass back the backpacks so that no student gets his or her own.

5.1.3 The Principle of Inclusion and Exclusion

The formula you have given in Problem 230 is often called **the principle of inclusion and exclusion** for unions of sets. The reason is the pattern in which the formula first adds (includes) all the sizes of the sets, then subtracts (excludes) all the sizes of the intersections of two sets, then adds (includes) all the sizes of the intersections of three sets, and so on. Notice that we haven't yet proved the principle. There are a variety of proofs. Perhaps one of the most straightforward (though not the most elegant) is an iductive proof that relies on the fact that

$$A_1 \cup A_2 \cup \cdots \cup A_n = (A_1 \cup A_2 \cup \cdots \cup A_{n-1}) \cup A_n$$

and the formula for the size of a union of two sets.

Problem 232. Give a proof of your formula for the principle of inclusion and exclusion. (h)

Problem 233. We get a more elegant proof if we ask for a picture enumerator for $A_1 \cup A_2 \cup \cdots \cup A_n$. so let us assume A is a set with a picture function P defined on it and that each set A_i is a subset of A.

(a) By thinking about how we got the formula for the size of a union, write down instead a conjecture for the picture enumerator of a union. You could use notation like $E_P(\bigcap_{i:i \in S} A_i)$ for the picture enumerator of the intersection of the sets A_i for i in a subset of S of $[n]$.

(b) If $x \in \bigcup_{i=1}^n A_i$, what is the coefficient for $P(x)$ in (the inclusion-exclusion side of) your formula for $E_P(\bigcup_{i=1}^n A_i)$? (h)

(c) If $x \notin \bigcup_{i=1}^n A_i$, what is the coefficient of $P(x)$ in (the inclusion-exclusion side of) your formula for $E_P(\bigcup_{i=1}^n A_i)$?

(d) How have you proved your conjecture for the picture enumerator of the union of the sets A_i?

(e) How can you get the formula for the principle of inclusion and exclusion from your formula for the picture enumerator of the union?

Problem 234. Frequently when we apply the principle of inclusion and exclusion, we will have a situation like that of part (d) of Problem 231.d. That is, we will have a set A and subsets $A_1, A_2, \ldots, A_n$ and we will want the size or the probability of the set of elements in A that are *not* in the union. This set is known as the **complement** of the union of the A_is in A, and is denoted by $A \setminus \bigcup_{i=1}^n A_i$, or if A is clear from context, by $\overline{\bigcup_{i=1}^n A_i}$. Give the fomula for $\overline{\bigcup_{i=1}^n A_i}$. The principle of inclusion and exclusion generall refers to both this formula and the one for the union.

We can find a very elegant way of writing the formula in Problem 234 if we let $\bigcap_{i:i \in \emptyset} A_i = A$. for this reason, if we have a family of subsets A_i of a set A, we define[1] $\bigcap_{i:i \in \emptyset} A_i = A$.

---

[1] For those interested in logic and set theory, given a family of subsets A_i of a set A, we define $\bigcap_{i:i \in S} A_i$ to be the set of all members x of A that are in A_i for all $i \in S$. (Note that this allows x to be in some other A_js as well.) Then if $S = \emptyset$, our intersection consists of all members x of A that satisfy the statement "if $i \in \emptyset$, then $x \in A_i$." But since the hypothesis of the "if-then" statement is false, the statement itself is true for all $x \in A$. Therefor $\bigcap_{i:i \in \emptyset} A_i = A$.

5.2 Application of Inclusion and Exclusion

5.2.1 Multisets with restricted numbers of elements

Problem 235. In how many ways may we distribute k identical apples to n children so that no child gets more than four apples? Compare your result with your result in Problem 197. (h)

5.2.2 The Ménage Problem

⇒ **Problem 236.** A group of n married couples comes to a group discussion session where they all sit around a round table. In how many ways can they sit so that no person is next to his or her spouse? (Note that two people of the same sex can sit next to each other.) (h)

⇒ * **Problem 237.** A group of n married couples comes to a group discussion session where they all sit around a round table. In how many ways can they sit so that no person is next to his or her spouse or a person of the same sex? This problem is called the **ménage problem**. (h)

5.2.3 Counting onto functions

- **Problem 238.** Given a function f from the k-element set K to the n-element set $[n]$, we say f is in the set A_i if $f(x) \neq i$ for every x in K. How many of these sets does an onto function belong to? What is the number of functions from a k-element set onto an n-element set?

⇒ **Problem 239.** Find a formula for the Stirling number (of the second kind) $S(k,n)$. (h)

Problem 240. If we roll a die eight times, we get a sequence of 8 numbers, the number of dots on top on the first roll, the number on the second roll, and so on.

   (a) What is the number of ways of rolling the die eight times so that each of the numbers one through six appears at least once in our sequence? To get a numerical answer, you will likely need a computer algebra package.

(b) What is the probability that we get a sequence in which all six numbers between one and six appear? To get a numerical answer, you will likely need a computer algebra package, programmable calculator, or spreadsheet.

(c) How many times do we have to roll the die to have probability at least one half that all six numbers appear in our sequence. To answer this question, you will likely need a computer algebra package, programmable calculator, or spreadsheet.

5.2.4 The chromatic polynomial of a graph

We defined a graph to consist of set V of elements called vertices and a set E of elements called edges such that each edge joins two vertices. A **coloring** of a graph by the elements of a set C (of colors) is an assignment of an element of C to each vertex of the graph; that is, a function from the vertex set V of the graph to C. A coloring is called **proper** if for each edge joining two distinct vertices[2], the two vertices it joins have different colors. You may have heard of the famous four color theorem of graph theory that says if a graph may be drawn in the plane so that no two edges cross (though they may touch at a vertex), then the graph has a proper coloring with four colors. Here we are interested in a different, though related, problem: namely, in how many ways may we properly color a graph (regardless of whether it can be drawn in the plane or not) using k or fewer colors? When we studied trees, we restricted ourselves to connected graphs. (Recall that a graph is connected if, for each pair of vertices, there is a walk between them.) Here, disconnected graphs will also be important to us. Given a graph which might or might not be connected, we partition its vertices into blocks called **connected components** as follows. For each vertex v we put all vertices connected to it by a walk into a block together. Clearly each vertex is in at least one block, because vertex v is connected to vertex v by the trivial walk consisting of the single vertex v and no edges. To have a partition, each vertex must be in one and only one block. To prove that we have defined a partition, suppose that vertex v is in the blocks B_1 and B_2. Then B_1 is the set of all vertices connected by walks to some vertex v_1 and B_2 is the set of all vertices connected by walks to some vertex v_2.

- **Problem 241.** (Relevant in Appendix C as well as this section.) Show that $B_1 = B_2$.

Since $B_1 = B_2$, these two sets are the same block, and thus all blocks containing v are identical, so v is in only one block. Thus we have a partition of the vertex set, and the blocks of the partition are the connected components of the graph. Notice that the connected components depend on the edge set of the graph. If we have a graph on the vertex set V with edge set E and another graph on the

---
[2]If a graph had a loop connecting a vertex to itself, that loop would connect a vertex to a vertex of the same color. It is because of this that we only consider edges with two distinct vertices in our definition.

vertex set V with edge set E', then these two graphs could have different connected components. It is traditional to use the Greek letter γ (gamma)[3] to stand for the number of connected components of a graph; in particular, $\gamma(V, E)$ stands for the number of connected components of the graph with vertex set V and edge set E. We are going to show how the principle of inclusion and exclusion may be used to compute the number of ways to properly color a graph using colors from a set C of c colors.

- **Problem 242.** Suppose we have a graph G with vertex set V and edge set E. Suppose F is a subset of E. Suppose we have a set C of c colors with which to color the vertices.

    (a) In terms of $\gamma(V, F)$, in how many ways may we color the vertices of G so that each edge in F connects two vertices of the same color? (h)

    (b) Given a coloring of G, for each edge e in E, let us consider the property that the endpoints of e are colored the same color. Let us call this property "property e." In this way each set of properties can be thought of as a subset of E. What set of properties does a proper coloring have?

    (c) Find a formula (which may involve summing over all subsets F of the edge set of the graph and using the number $\gamma(V, F)$ of connected components of the graph with vertex set V and edge set F) for the number of proper colorings of G using colors in the set C. (h)

The formula you found in Problem 242.c is a formula that involves powers of c, and so it is a polynomial function of c. Thus it is called the **chromatic polynomial** of the graph G. Since we like to think about polynomials as having a variable x and we like to think of c as standing for some constant, people often use x as the notation for the number of colors we are using to color G. Frequently people will use $\chi_G(x)$ to stand for the number of way to color G with x colors, and call $\chi_G(x)$ the **chromatic polynomial** of G.

5.3 Deletion-Contraction and the Chromatic Polynomial

⇒ **Problem 243.** In Chapter 2 we introduced the deletion-contraction recurrence for counting spanning trees of a graph. Figure out how the chromatic polynomial of a graph is related to those resulting from deletion of an edge e and from contraction of that same edge e. Try to find a recurrence like the one for counting spanning trees that expresses the chromatic polynomial of a graph in terms of the chromatic polynomials of $G - e$ and G/e for an

---

[3] The greek letter gamma is pronounced gam-uh, where gam rhymes with ham.

arbitrary edge e. Use this recurrence to give another proof that the number of ways to color a graph with x colors is a polynomial function of x. (h)

Problem 244. Use the deletion-contraction recurrence to compute the chromatic polynomial of the graph in Figure 5.3.1. (You can simplify your computations by thinking about the effect on the chromatic polynomial of deleting an edge that is a loop, or deleting one of several edges between the same two vertices.)

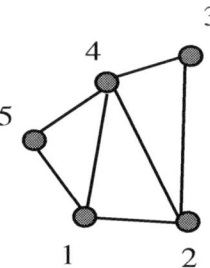

Figure 5.3.1: A graph.

⇒ **Problem 245.**

(a) In how many ways may you properly color the vertices of a path on n vertices with x colors? Describe any dependence of the chromatic polynomial of a path on the number of vertices.

(b) (Not tremendously hard.) In how many ways may you properly color the vertices of a cycle on n vertices with x colors? Describe any dependence of the chromatic polynomial of a cycle on the number of vertices.

Problem 246. In how many ways may you properly color the vertices of a tree on n vertices with x colors? (h)

⇒ **Problem 247.** What do you observe about the signs of the coefficients of the chromatic polynomial of the graph in Figure 5.3.1? What about the signs of the coefficients of the chromatic polynomial of a path? Of a cycle? Of a tree? Make a conjecture about the signs of the coefficients of a chromatic polynomial and prove it.

5.4 Supplementary Problems

1. Each person attending a party has been asked to bring a prize. The person planning the party has arranged to give out exactly as many prizes as there are guests, but any person may win any number of prizes. If there are n guests, in how many ways may the prizes be given out so that nobody gets the prize that he or she brought?

2. There are m students attending a seminar in a room with n seats. The seminar is a long one, and in the middle the group takes a break. In how many ways may the students return to the room and sit down so that nobody is in the same seat as before?

3. What is the number of ways to pass out k pieces of candy from an unlimited supply of identical candy to n children (where n is fixed) so that each child gets between three and six pieces of candy (inclusive)? If you have done Problem 1 of Supplementary Problems 4.4, compare your answer in that problem with your answer in this one.

⇒ **4.** In how many ways may k distinct books be arranged on n shelves so that no shelf gets more than m books?

⇒ **5.** Suppose that n children join hands in a circle for a game at nursery school. The game involves everyone falling down (and letting go). In how many ways may they join hands in a circle again so that nobody is to the right of the same child that was previously to his or her right?

⇒ ∗ **6.** Suppose that n people link arms in a folk-dance and dance in a circle. Later on they let go and dance some more, after which they link arms in a circle again. In how many ways can they link arms the second time so that no-one is next to a person with whom he or she linked arms before.

⇒ ∗ **7.** (A challenge; the author has not tried to solve this one!) Redo Problem 6 in the case that there are n men and n women and when people arrange themselves in a circle they do so alternating gender.

⇒ **8.** Suppose we take two graphs G_1 and G_2 with disjoint vertex sets, we choose one vertex on each graph, and connect these two graphs by an edge e to get a graph G_{12}. How does the chromatic polynomial of G_{12} relate to those of G_1 and G_2?

Chapter 6

Groups acting on sets

6.1 Permutation Groups

Until now we have thought of permutations mostly as ways of listing the elements of a set. In this chapter we will find it very useful to think of permutations as functions. This will help us in using permutations to solve enumeration problems that cannot be solved by the quotient principle because they involve counting the blocks of a partition in which the blocks don't have the same size. We begin by studying the kinds of permutations that arise in situations where we have used the quotient principle in the past.

6.1.1 The rotations of a square

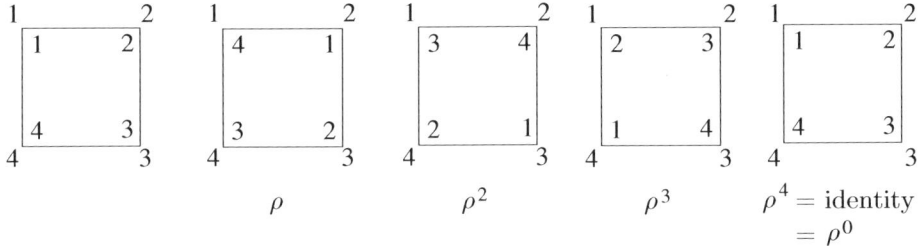

Figure 6.1.1: The four possible results of rotating a square and maintaining its position.

In Figure 6.1.1 we show a square with its four vertices labelled 1, 2, 3, and 4. We have also labeled the spot in the plane where each of these vertices falls with the same label. Then we have shown the effect of rotating the square clockwise through 90, 180, 270, and 360 degrees (which is the same as rotating through 0 degrees). Underneath each of the rotated squares we have named the function that carries out the rotation. We use ρ, the Greek letter pronounced "row," to stand for a 90 degree clockwise rotation. We use ρ^2 to stand for two 90 degree rotations, and so on. We can think of the function ρ as a function on the four element set[1] $\{1, 2, 3, 4\}$. In particular, for any function φ (the Greek letter phi, usually pronounced "fee," but sometimes "fie") from the plane back to itself that may move the square around but otherwise leaves it in the same place, we let $\varphi(i)$ be the label of the place where vertex previously in position i is now. Thus $\rho(1) = 2$, $\rho(2) = 3$, $\rho(3) = 4$ and $\rho(4) = 1$. Notice that ρ is a permutation on the set $\{1, 2, 3, 4\}$.

- **Problem 248.** The composition $f \circ g$ of two functions f and g is defined by $f \circ g(x) = f(g(x))$. Is ρ^3 the composition of ρ and ρ^2? Does the answer depend on the order in which we write ρ and ρ^2? How is ρ^2 related to ρ?

- **Problem 249.** Is the composition of two permutations always a permutation?

In Problem 248 you see that we can think of $\rho^2 \circ \rho$ as the result of first rotating by 90 degrees and then by another 180 degrees. In other words, the composition of two rotations is the same thing as first doing one and then doing the other. Of course there is nothing special about 90 degrees and 180 degrees. As long as we first do one rotation through a multiple of 90 degrees and then another rotation through a multiple of 90 degrees, the composition of these rotations is a rotation through a multiple of 90 degrees.

If we first rotate by 90 degrees and then by 270 degrees then we have rotated by 360 degrees, which does nothing visible to the square. Thus we say that ρ^4 is the "identity function." In general the **identity function** on a set S, denoted by ι (the Greek letter iota, pronounced eye-oh-ta) is the function that takes each element of the set to itself. In symbols, $\iota(x) = x$ for every x in S. Of course the identity function on a set is a permutation of that set.

6.1.2 Groups of Permutations

- **Problem 250.** For any function φ from a set S to itself, we define φ^n (for nonnegative integers n) inductively by $\varphi^0 = \iota$ and $\varphi^n = \varphi^{n-1} \circ \varphi$ for every positive integer n. If φ is a permutation, is φ^n a permutation? Based on your experience with previous inductive proofs, what do you expect $\varphi^n \circ \varphi^m$ to be? What do you expect $(\varphi^m)^n$ to be? There is no need to prove these last two answers are correct, for you have, in effect, already done so in Chapter 2.

- **Problem 251.** If we perform the composition $\iota \circ \varphi$ for any function φ from S to S, what function do we get? What if we perform the composition $\varphi \circ \iota$?

---

[1] What we are doing is restricting the rotation ρ to the set $\{1, 2, 3, 4\}$.

What you have observed about iota in Problem 251 is called the **identity property** of iota. In the context of permutations, people usually call the function ι "the identity" rather than calling it "iota."

Since rotating first by 90 degrees and then by 270 degrees has the same effect as doing nothing, we can think of the 270 degree rotation as undoing what the 90 degree rotation does. For this reason we say that in the rotations of the square, ρ^3 is the "inverse" of ρ. In general, a function $\varphi : T \to S$ is called an **inverse** of a function $\sigma : S \to T$ (the lower case Greek letter sigma) if $\varphi \circ \sigma = \sigma \circ \varphi = \iota$. For a slower introduction to inverses and practice with them, see Section A.1.3 in Appendix A. Since a permutation is a bijection, it has a unique inverse, as in Section A.1.3. And since the inverse of a bijection is a bijection (again, as in the Appendix), the inverse of a permutation is a permutation.

We use φ^{-1} to denote the inverse of the permutation φ. We've seen that the rotations of the square are functions that return the square to its original position but may move the vertices to different places. In this way we create permutations of the vertices of the square. We've observed three important properties of these permutations.

- (Identity Property) These permutations include the identity permutation.

- (Inverse Property) Whenever these permutations include φ, they also include φ^{-1}.

- (Closure Property) Whenever these permutations include φ and σ, they also include $\varphi \circ \sigma$.

A set of permutations with these three properties is called a **permutation group**[2] or a group of permutations. We call the group of permutations corresponding to rotations of the square the **rotation group** of the square. There is a similar rotation group with n elements for any regular n-gon.

- **Problem 252.** If $f : S \to T$, $g : T \to X$, and $h : X \to Y$, is $h \circ (g \circ f) = (h \circ g) \circ f$? What does this say about the status of the *associative law*

$$\rho \circ (\sigma \circ \varphi) = (\rho \circ \sigma) \circ \varphi$$

in a group of permutations?

- **Problem 253.**

    (a) How should we define φ^{-n} for an element φ of a permutation group? (h)

    (b) Will the two standard rules for exponents

    $$a^m a^n = a^{m+n} \text{ and } (a^m)^n = a^{mn}$$

---

[2]The concept of a permutation group is a special case of the concept of a **group** that one studies in abstract algebra. When we refer to a group in what follows, if you know what groups are in the more abstract sense, you may use the word in this way. If you do not know about groups in this more abstract sense, then you may assume we mean permutation group when we say group.

still hold if one or more of the exponents may be negative?

(c) What would we have to prove to show that the rules still hold?

(d) If the rules hold, give enough of the proof to show that you know how to do it; otherwise give a counterexample.

- **Problem 254.** If a finite set of permutations satisfies the closure property is it a permutation group? (h)

- **Problem 255.** There are three-dimensional geometric motions of the square that return it to its original position but move some of the vertices to other positions. For example, if we flip the square around a diagonal, most of it moves out of the plane during the flip, but the square ends up in the same place. Draw a figure like Figure 6.1.1 that shows all the possible results of such motions, including the ones shown in Figure 6.1.1. Do the corresponding permutations form a group?

Problem 256. Let σ and φ be permutations.

(a) Why must $\sigma \circ \varphi$ have an inverse?

(b) Is $(\sigma \circ \varphi)^{-1} = \sigma^{-1} \varphi^{-1}$? (Prove or give a counter-example.) (h)

(c) Is $(\sigma \circ \varphi)^{-1} = \varphi^{-1} \sigma^{-1}$? (Prove or give a counter-example.)

- **Problem 257.** Explain why the set of all permutations of four elements is a permutation group. How many elements does this group have? This group is called the **symmetric group on four letters** and is denoted by S_4.

6.1.3 The symmetric group

In general, the set of all permutations of an n-element set is a group. It is called the **symmetric group on n letters**. We don't have nice geometric descriptions (like rotations) for all its elements, and it would be inconvenient to have to write down something like "Let $\sigma(1) = 3$, $\sigma(2) = 1$, $\sigma(3) = 4$, and $\sigma(4) = 1$" each time we need to introduce a new permutation. We introduce a new notation for permutations that allows us to denote them *reasonably* compactly and compose them reasonably quickly. If σ is the permutation of $\{1, 2, 3, 4\}$ given by $\sigma(1) = 3$, $\sigma(2) = 1$, $\sigma(3) = 4$

and $\sigma(4) = 2$, we write
$$\sigma = \begin{pmatrix} 1 & 2 & 3 & 4 \\ 3 & 1 & 4 & 2 \end{pmatrix}.$$

We call this notation the **two row notation** for permutations. In the two row notation for a permutation of $\{a_1, a_2, \ldots, a_n\}$, we write the numbers a_1 through a_n in a one row and we write $\sigma(a_1)$ through $\sigma(a_n)$ in a row right below, enclosing both rows in parentheses. Notice that

$$\begin{pmatrix} 1 & 2 & 3 & 4 \\ 3 & 1 & 4 & 2 \end{pmatrix} = \begin{pmatrix} 2 & 1 & 4 & 3 \\ 1 & 3 & 2 & 4 \end{pmatrix},$$

although the second ordering of the columns is rarely used.

If φ is given by
$$\varphi = \begin{pmatrix} 1 & 2 & 3 & 4 \\ 4 & 1 & 2 & 3 \end{pmatrix},$$

then, by applying the definition of composition of functions, we may compute $\sigma \circ \varphi$ as shown in Figure 6.1.2.

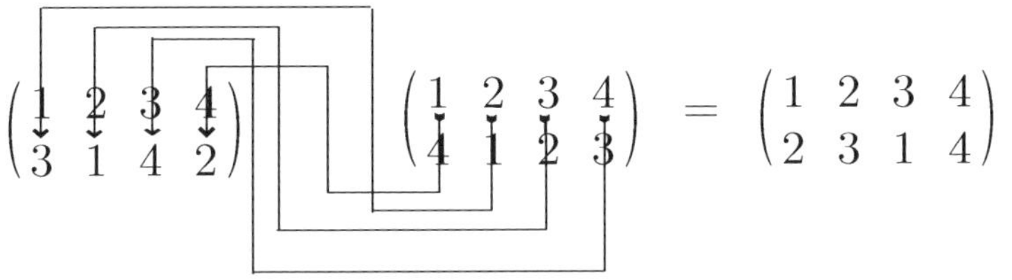

Figure 6.1.2: How to multiply permutations in two-row notation.

We don't normally put the circle between two permutations in two row notation when we are composing them, and refer to the operation as multiplying the permutations, or as the product of the permutations. To see how Figure 6.1.2 illustrates composition, notice that the arrow starting at 1 in φ goes to 4. Then from the 4 in φ it goes to the 4 in σ and then to 2. This illustrates that $\varphi(1) = 4$ and $\sigma(4) = 2$, so that $\sigma(\varphi(1)) = 2$.

Problem 258. For practice, compute $\begin{pmatrix} 1 & 2 & 3 & 4 & 5 \\ 3 & 4 & 1 & 5 & 2 \end{pmatrix} \begin{pmatrix} 1 & 2 & 3 & 4 & 5 \\ 4 & 3 & 5 & 1 & 2 \end{pmatrix}.$

6.1.4 The dihedral group

We found four permutations that correspond to rotations of the square. In Problem 255 you found four permutations that correspond to flips of the square in space. One flip fixes the vertices in the places labeled 1 and 3 and interchanges the vertices in the places labeled 2 and 4. Let us denote it by $\varphi_{1|3}$. One flip fixes the vertices in the positions labeled 2 and 4 and interchanges those in the positions labeled 1 and 3. Let us denote it by $\varphi 2|4$. One flip interchanges the vertices in the places labeled 1 and 2 and also interchanges those in the places labeled 3 and 4.

Let us denote it by $\varphi_{12|34}$. The fourth flip interchanges the vertices in the places labeled 1 and 4 and interchanges those in the places labeled 2 and 3. Let us denote it by $\varphi_{14|23}$. Notice that $\varphi_{1|3}$ is a permutation that takes the vertex in place 1 to the vertex in place 1 and the vertex in place 3 to the vertex in place 3, while $\varphi_{12|34}$ is a permutation that takes the edge between places 1 and 2 to the edge between places 2 and 1 (which is the same edge) and takes the edge between places 3 and 4 to the edge between places 4 and 3 (which is the same edge). This should help to explain the similarity in the notation for the two different kinds of flips.

- **Problem 259.** Write down the two-row notation for ρ^3, $\varphi_{2|4}$, $\varphi_{12|34}$ and $\varphi_{2|4} \circ \varphi_{12|34}$. Remember that $\sigma(i)$ stands for the position where the vertex that originated in position i is after we apply σ.

Problem 260. (You may have already done this problem in Problem 255, in which case you need not do it again!) In Problem 255, if a rigid motion of three-dimensional space returns the square to its original position, in how many places can vertex number one land? Once the location of vertex number one is decided, how many possible locations are there for vertex two? Once the locations of vertex one and vertex two are decided, how many locations are there for vertex three? Answer the same question for vertex four. What does this say about the relationship between the four rotations and four flips described above and the permutations you described in Problem 255?

The four rotations and four flips of the square described before Problem 259 form a group called the dihedral group of the square. Sometimes the group is denoted D_8 because it has eight elements, and sometimes the group is denoted by D_4 because it deals with four vertices! Let us agree to use the notation D_4 for the dihedral group of the square. There is a similar dihedral group, denoted by D_n, of all the rigid motions of three-dimensional space that return a regular n-gon to its original position (but might put the vertices in different places.)

Problem 261. Another view of the dihedral group of the square is that it is the group of all distance preserving functions, also called **isometries**, from a square to itself. Notice that an isometry must be a bijection. Any rigid motion of the square preserves the distances between all points of the square. However, it is conceivable that there might be some isometries that do not arise from rigid motions. (We will see some later on in the case of a cube.) Show that there are exactly eight isometires (distance preserving functions) from a square to itself. (h)

Problem 262. How many elements does the group D_n have? Prove that you are correct.

Problem 263. In Figure 6.1.3 we show a cube with the positions of its vertices and faces labeled. As with motions of the square, we let $\varphi(x)$ be the label of the place where vertex previously in position x is now.

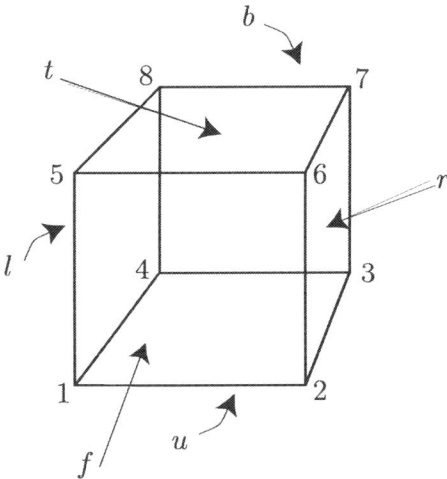

Figure 6.1.3: A cube with the positions of its vertices and faces labelled. The curved arrows point to the positions that are blocked by the cube.

(a) Write in two row notation the permutation ρ of the vertices that corresponds to rotating the cube 90 degrees around a vertical axis through the faces t (for top) and u (for underneath). (Rotate in a right-handed fashion around this axis, meaning that vertex 6 goes to the back and vertex 8 comes to the front.)

(b) Write in two row notation the permutation φ that rotates the cube 120 degrees around the diagonal from vertex 1 to vertex 7 and carries vertex 8 to vertex 6.

(c) Compute the two row notation for $\rho \circ \varphi$

(d) Is the permutation $\rho \circ \varphi$ a rotation of the cube around some axis? If so, say what the axis is and how many degrees we rotate around the axis. If $\rho \circ \varphi$ is not a rotation, give a geometic description of it.

⇒ · **Problem 264.** How many permutations are in the group R? R is sometimes called the "rotation group" of the cube. Can you justify this? (h)

Problem 265. As with a two-dimensional figure, it is possible to talk about isometries of a three-dimensional figure. These are distance preserving functions from the figure to itself. The function that reflects the cube in Figure 6.1.3 through a plane halfway between the bottom face and top face exchanges the vertices 1 and 5, 2 and 6, 3 and 7, and 4 and 8 of the cube. This function preserves distances between points in the cube. However, it cannot be achieved by a rigid motion of the cube because a rigid motion that takes vertex 1 to vertex 5, vertex 2 to vertex 6, vertex 3 to vertex 7, and vertex 4 to vertex 8 would not return the cube to its original location; rather it would put the bottom of the cube where its top previously was and would put the rest of the cube above that square rather than below it.

(a) How many elements are there in the group of permutations of [8] that correspond to isometries of the cube? (h)

(b) Is every permutation of [8] that corresponds to an isometry either a rotation or a reflection? (h)

6.1.5 Group tables (Optional)

We can always figure out the composition of two permutations of the same set by using the definition of composition, but if we are going to work with a given permutation group again and again, it is worth making the computations once and recording them in a table. For example the group of rotations of the square may be represented as in Table 6.1.4. We list the elements of our group, with the identity first, across the top of the table and down the left side of the table, using the same order both times. Then in the row labeled by the group element σ and the column labelled by the group element φ, we write the composition $\sigma \circ \varphi$, expressed in terms of the elements we have listed on the top and on the left side. Since a group of permutations is closed under composition, the result $\sigma \circ \varphi$ will always be expressible as one of these elements.

$\circ$	ι	ρ	ρ^2	ρ^3
ι	ι	ρ	ρ^2	ρ^3
ρ	ρ	ρ^2	ρ^3	ι
ρ^2	ρ^2	ρ^3	ι	ρ
ρ^3	ρ^3	ι	ρ	ρ^2

Table 6.1.4: The group table for the rotations of a square.

Problem 266. In Table 6.1.4, all the entries in a row (not including the first entry, the one to the left of the line) are different. Will this be true in any group table for a permutation group? Why or why not? Also in Table 6.1.4 all the entries in a column (not including the first entry, the one above the line) are different. Will this be true in any group table for a permutation group? Why or why not?

Problem 267. In Table 6.1.4, every element of the group appears in every row (even if you don't include the first element, the one before the line). Will this be true in any group table for a permutation group? Why or why not? Also in Table 6.1.4 every element of the group appears in every column (even if you don't include the first entry, the one before the line). Will this be true in any group table for a permutation group? Why or why not?

- **Problem 268.** Write down the group table for the dihedral group D_4. Use the φ notation described above to denote the flips. (Hints: Part of the table has already been written down. Will you need to think hard to write down the last row? Will you need to think hard to write down the last column? When you multiply a product like $\varphi_{1|3} \circ \rho$ remember that we defined $\varphi_{1|3}$ to be the flip that fixes the vertex in position 1 and the vertex in position 3, *not* the one that fixes the vertex on the square labelled 1 and the vertex on the square labelled 3.)

You may notice that the associative law, the identity property, and the inverse property are three of the most important rules that we use in regrouping parentheses in algebraic expressions when solving equations. There is one property we have not yet mentioned, the **commutative law** which would say that $\sigma \circ \varphi = \varphi \circ \sigma$. It is easy to see from the group table of R_4 that it satisfies the commutative law.

Problem 269. Does the commutative law hold in all permutation groups?

6.1.6 Subgroups

We have seen that the dihedral group D_4 contains a copy of the group of rotations of the square. When one group G of permutations of a set S is a subset of another group G' of permutations of S, we say that G is a **subgroup** of G'.

- **Problem 270.** Find all subgroups of the group D_4. (h)

Problem 271. Can you find subgroups of the symmetric group S_4 with two elements? Three elements? Four elements? Six elements? (For each positive answer, describe a subgroup. For each negative answer, explain why not.)

6.1.7 The cycle structure of a permutation

There is an even more efficient way to write down permutations. Notice that the product in Figure 6.1.2 is $\begin{pmatrix} 1 & 2 & 3 & 4 \\ 2 & 3 & 1 & 4 \end{pmatrix}$. We have drawn the directed graph of this permutation in Figure 6.1.5.

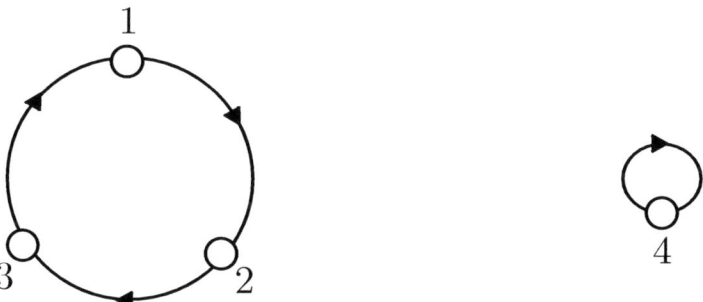

Figure 6.1.5: The directed graph of a permutation.

You see that the graph has two directed cycles, the rather trivial one with vertex 4 pointing to itself, and the nontrivial one with vertex 1 pointing to vertex 2 pointing to vertex 3 which points back to vertex 1. A permutation is called a **cycle** if its digraph consists of exactly one cycle. Thus $\begin{pmatrix} 1 & 2 & 3 \\ 2 & 3 & 1 \end{pmatrix}$ is a cycle but $\begin{pmatrix} 1 & 2 & 3 & 4 \\ 2 & 3 & 1 & 4 \end{pmatrix}$ is not a cycle by our definition. We write (1 2 3) or (2 3 1) or (3 1 2) to stand for the cycle $\sigma = \begin{pmatrix} 1 & 2 & 3 \\ 2 & 3 & 1 \end{pmatrix}$.

We can describe cycles in another way as well. A **cycle** of the permutation σ is a list $(i\ \sigma(i)\ \sigma^2(i)\ \ldots\ \sigma^n(i))$ that does not have repeated elements while the list $(i\ \sigma(i)\ \sigma^2(i)\ \ldots\ \sigma^n(i))\ \sigma^{n+1}(i))$ does have repeated elements.

> **Problem 272.** If the list $(i\ \sigma(i)\ \sigma^2(i)\ \ldots\ \sigma^n(i))$ does not have repeated elements but the list $(i\ \sigma(i)\ \sigma^2(i)\ \ldots\ \sigma^n(i)\ \sigma^{n+1}(i))$ does have repeated elements, then what is $\sigma^{n+1}(i)$? (h)

We say $\sigma^j(i)$ is an **element** of the cycle $(i\ \sigma(i)\ \sigma^2(i)\ \ldots\ \sigma^n(i))$. Notice that the case $j = 0$ means i is an element of the cycle. Notice also that if $j > n$, $\sigma^j(i) = \sigma^{j-n-1}(i)$, so the distinct elements of the cycle are i, $\sigma(i)$, $\sigma^2(i)$, through $\sigma^n(i)$. We think of the cycle $(i\ \sigma(i)\ \sigma^2(i)\ \ldots\ \sigma^n(i))$ as representing the permutation σ restricted to the set of elements of the cycle. We say that the cycles $(i\ \sigma(i)\ \sigma^2(i)\ \ldots\ \sigma^n(i))$ and $(j\ \sigma(j)\ \sigma^2(j)\ \ldots\ \sigma^n(j))$ are **equivalent** if there is an integer k such that $j = \sigma^k(i)$.

- **Problem 273.** Find the cycles of the permutations ρ, $\varphi_{1|3}$ and $\varphi_{12|34}$ in the group D_4.

Problem 274. Find the cycles of the permutation

$$\begin{pmatrix} 1 & 2 & 3 & 4 & 5 & 6 & 7 & 8 & 9 \\ 3 & 4 & 6 & 2 & 9 & 7 & 1 & 5 & 8 \end{pmatrix}.$$

Problem 275. If two cycles of σ have an element in common, what can we say about them?

Problem 275 leads almost immediately to the following theorem.

Theorem 6.1.6. *for each permutation σ of a set S, there is a unique partition of S each of whose blocks is the set of elements of a cycle of σ.*

More informally, we may say that every permutation partitions its domain into disjoint cycles. We call the set of cycles of a permutation the **cycle decomposition** of the permutation. Since the cycles of a permutation σ tell us $\sigma(x)$ for every x in the domain of σ, the cycle decomposition of a permutation completely determines the permutation. Using our informal language, we can express this idea in the following corollary to Theorem 6.1.6.

Corollary 6.1.7. *Every partition of a set S into cycles determins a unique permutation of S.*

Problem 276. Prove Theorem 6.1.6.

In Problems 273 and Problem 274 you found the cycle decomposition of typical elements of the group D_4 and of the permutation

$$\begin{pmatrix} 1 & 2 & 3 & 4 & 5 & 6 & 7 & 8 & 9 \\ 3 & 4 & 6 & 2 & 9 & 7 & 1 & 5 & 8 \end{pmatrix}$$

The group of all rotations of the square is simply the set of the four powers of the cycle $\rho = (1\ 2\ 3\ 4)$. for this reason it is called a **cyclic group**[3] and is often denoted by C_4. Similarly, the rotation group of an n-gon is usually denoted C_n.

⇒ **Problem 277.** Write a recurrence for the number $c(k, n)$ for the number of permutations of $[k]$ that have exactly n cycles, including 1-cycles. Use it to write a table of $c(k, n)$ for k between 1 and 7 inclusive. Can you find a relationship between $c(k, n)$ and any of the other families of special numbers such as binomial coefficients, Stirling numbers, Lah numbers, etc. we have studied? If you find such a relationship, prove you are right. (h)

---

[3]The phrace cyclic group applies in a more general (but similar) situation. Namely the set of all powers of any member of a group is called a cyclic group.

⇒ · **Problem 278.** (Relevant to Appendix C.) A permutation σ is called an **involution** if $\sigma^2 = \iota$. When you write an involution as a product of disjoint cycles, what is special about the cycles?

6.2 Groups Acting on Sets

We defined the rotation group R_4 and the dihedral group D_4 as groups of permutations of the vertices of a square. These permutations represent rigid motions of the square in the plane and in three dimensional space respectively. The square has geometric features of interest other than its vertices; for example its diagonals, or its edges. Any geometric motion of the square that returns it to its original position takes each diagonal to a possibly different diagonal, and takes each edge to a possibly different edge. In Figure 6.2.1 we show the results on the sides and diagonals of the rotations of a square. The rotation group permutes the sides of the square and permutes the diagonals of the square as it rotates the square. Thus, we say the rotation group "acts" on the sides and diagonals of the square.

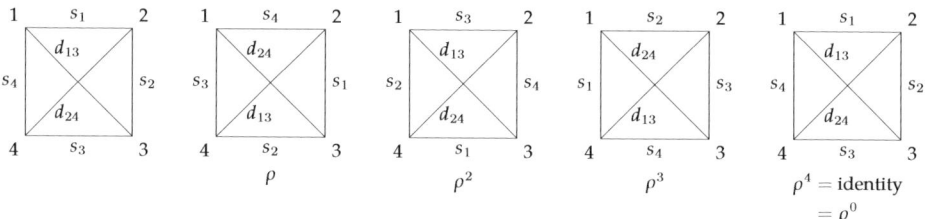

Figure 6.2.1: The results on the sides and diagonals of rotating the square

Problem 279.

(a) Write down the two-line notation for the permutation $\overline{\rho}$ that a 90 degree rotation does to the sides of the square.

(b) Write down the two-line notation for the permutation $\overline{\rho^2}$ that a 180 degree rotation does to the sides of the square.

(c) Is $\overline{\rho^2} = \overline{\rho} \circ \overline{\rho}$? Why or why not?

(d) Write down the two-line notation for the permutation $\widehat{\rho}$ that a 90 degree rotation does to the diagonals d_{13} and d_{24} of the square.

(e) Write down the two-line notation for the permutation $\widehat{\rho^2}$ that a 180 degree rotation does to the diagonals d_{13} and d_{24} of the square.

(f) Is $\widehat{\rho^2} = \widehat{\rho} \circ \widehat{\rho}$? Why or why not? What familiar permutation is $\widehat{\rho^2}$ in this case?

We have seen that the fact that we have defined a permutation group as the permutations of some specific set doesn't preclude us from thinking of the elements of that group as permuting the elements of some other set as well. In order to keep track of which permutations of which set we are using to define our group and which other set is being permuted as well, we introduce some new language and notation. We are going to say that the group D_4 "acts" on the edges and diagonals of a square and the group R of permutations of the vertices of a cube that arise from rigid motions of the cube "acts" on the edges, faces, diagonals, etc. of the cube.

- **Problem 280.** In Figure 6.1.3 we show a cube with the positions of its vertices and faces labeled. As with motions of the square, we let we let $\varphi(x)$ be the label of the place where vertex previously in position x is now.

  (a) In Problem 263 we wrote in two row notation the permutation ρ of the vertices that corresponds to rotating the cube 90 degrees around a vertical axis through the faces t (for top) and u (for underneath). (We rotated in a right-handed fashion around this axis, meaning that vertex 6 goes to the back and vertex 8 comes to the front.) Write in two row notation the permutation $\overline{\rho}$ of the faces that corresponds to this member ρ of R.

  (b) In Problem 263 we wrote in two row notation the permutation φ that rotates the cube 120 degrees around the diagonal from vertex 1 to vertex 7 and carries vertex 8 to vertex 6. Write in two row notation the $\overline{\varphi}$ of the faces that corresponds to this member of R.

  (c) In Problem 263 we computed the two row notation for $\rho \circ \varphi$. Now compute the two row notation for $\overline{\rho} \circ \overline{\varphi}$ ($\overline{\rho}$ was defined in Part 280.a), and write in two row notation the permutation $\overline{\rho \circ \varphi}$ of the faces that corresponds to the action of the permutation $\rho \circ \varphi$ on the faces of the cube. (For this question it helps to think geometrically about what motion of the cube is carried out by $\rho \circ \varphi$.) What do you observe about $\overline{\rho \circ \varphi}$ and $\overline{\rho} \circ \overline{\varphi}$?

We say that a permutation group G **acts** on a set S if, for each member σ of G there is a permutation $\overline{\sigma}$ of S such that

$$\overline{\sigma \circ \varphi} = \overline{\sigma} \circ \overline{\varphi}$$

for every member σ and φ of G. In Problem 280.c you saw one example of this condition. If we think intuitively of ρ and φ as motions in space, then following the action of φ by the action of ρ does give us the action of $\rho \circ \varphi$. We can also reason directly with the permutations in the group R of rigid motions (rotations) of the cube to show that R acts on the faces of the cube.

Problem 281. Show that a group G of permutations of a set S acts on S with $\overline{\varphi} = \varphi$ for all φ in G.

- **Problem 282.** The group D_4 is a group of permutations of $\{1,2,3,4\}$ as in Problem 255. We are going to show in this problem how this group acts on the two-element subsets of $\{1,2,3,4\}$. In Problem 287 we will see a natural geometric interpretation of this action. In particular, for each two-element subset $\{i,j\}$ of $\{1,2,3,4\}$ and each member σ of D_4 we define $\overline{\sigma}(\{i,j\}) = \{\sigma(i), \sigma(j)\}$. Show that with this definition of $\overline{\sigma}$, the group D_4 acts on the two-element subsets of $\{1,2,3,4\}$.

- **Problem 283.** Suppose that σ and φ are permutations in the group R of rigid motions of the cube. We have argued already that each rigid motion sends a face to a face. Thus σ and φ both send the vertices on one face to the vertices on another face. Let $\{h, i, j, k\}$ be the set of labels next to the vertices on a face F.

    (a) What are the vertices of the face F' that F is sent to by φ?

    (b) What are the vertices of the face F'' that F' is sent to by σ?

    (c) What are the vertices of the face F''' that F is sent to by $\sigma \circ \varphi$?

    (d) How have you just shown that the group R acts on the faces?

6.2.1 Groups acting on colorings of sets

Recall that when you were asked in Problem 45 to find the number of ways to place two red beads and two blue beads at the corners of a square free to move in three-dimensional space, you were not able to apply the quotient principle to answer the question. Instead you had to see that you could divide the set of six lists of two Rs and two Bs into two sets, one of size two in which the Rs and Bs alternated and one of size four in which the two reds (and therefore the two blues) would be side-by-side on the square. Saying that the square is free to move in space is equivalent to saying that two arrangements of beads on the square are equivalent if a member of the dihedral group carries one arrangement to the other. Thus an important ingredient in the analysis of such problems will be how a group can act on colorings of a set of vertices. We can describe the coloring of the square in Figure 6.2.2 as the function f with

$$f(1) = R, \ f(2) = R, \ f(3) = B, \text{ and } f(4) = B,$$

but it is more compact and turns out to be more suggestive to represent the coloring in Figure 6.2.2 as the set of ordered pairs

$$(1, R), (2, R), (3, B), (4, B) \tag{6.1}$$

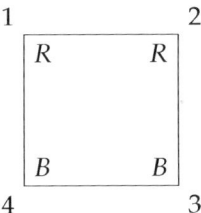

Figure 6.2.2: The colored square with coloring $\{(1,R),(2,R),(3,B),(4,B)\}$

This gives us an explicity list of which colors are assigned to which vertex.[4] Then if we rotate the square through 90 degrees, we see that the set of ordered pairs becomes
$$\{(\rho(1),R),(\rho(2),R),(\rho(3),B),(\rho(4),B)\} \tag{6.2}$$
which is the same as
$$\{(2,R),(3,R),(4,B),(1,B)\}.$$
Or, in a more natural order,
$$\{(1,B),(2,R),(3,R),(4,B)\}. \tag{6.3}$$

The reordering we did in (6.3) suggests yet another simplification of notation. So long as we know we that the first elements of our pairs are labeled by the members of $[n]$ for some integer n and we are listing our pairs in increasing order by the first component, we can denote the coloring
$$\{(1,B),(2,R),(3,R),(4,B)\}$$
by $BRRB$. In the case where we have numbered the elements of the set S we are coloring, we will call this list of colors of the elements of S in order the **standard notation** for the coloring. We will call the ordering used in (6.3) the **standard ordering** of the coloring.

Thus we have three natural ways to represent a coloring of a set: as a function, as a set of ordered pairs, and as a list. Different representations are useful for different things. For example, the representation by ordered pairs will provide a natural way to define the action of a group on colorings of a set. Given a coloring as a function f, we denote the set of ordered pairs
$$\{(x,f(x)) \mid x \in S\},$$
suggestively as (S,f) for short. We use $f(1)f(2)\cdots f(n)$ to stand for a particular coloring (S,f) in the standard notation.

Problem 284. Suppose now that instead of coloring the vertices of a square, we color its edges. We will use the shorthand 12, 23, 34, and 41 to stand for the edges of the square between vertex 1 and vertex 2, vertex 2 and vertex

---

[4] The reader who has studied Appendix A will recognize that this set of ordered pairs is the relation of the function f, but we won't need to make any specific references to the idea of a relation in what follows.

3, and so on. Then a coloring of the edges with 12 red, 23 blue, 34 red and 41 blue can be represented as

$$\{(12, R), (23, B), (34, R), (41, B)\}. \tag{6.4}$$

If ρ is the rotation through 90 degrees, then we have a permutation $\overline{\rho}$ acting on its edges. This permutation acts on the colorings to give us a permutation $\overline{\overline{\rho}}$ of the set of colorings.

(a) What is $\overline{\overline{\rho}}$ of the coloring in (6.4)?

(b) What is $\overline{\overline{\rho^2}}$ of the coloring in (6.4)?

If G is a group that acts on the set S, we define the **action of G on the colorings** (S, f) by by

$$\overline{\overline{\sigma}}((S, f)) = \overline{\overline{\sigma}}\left(\{(x, f(x)) \mid x \in S\}\right) = \{(\overline{\sigma}(x), f(x)) \mid x \in S\}.. \tag{6.5}$$

We have two bars over σ because σ is a permutation of one set that gives us a permutation $\overline{\sigma}$ of a second set, and then $\overline{\sigma}$ acts to give a permutation $\overline{\overline{\sigma}}$ of a thid set, the set of colorings. For example, suppose we want to analyze colorings of the faces of a cube under the action of the rotation group of the cube as we have defined it on the vertices. Each vertex-permutation σ in the group gives a permutation $\overline{\sigma}$ of the faces of the cube. Then each permutation $\overline{\sigma}$ of the faces gives us a permutation $\overline{\overline{\sigma}}$ of the colorings of the faces.

In the special case that G is a group of permutations of S rather than a group acting on S, Equation (6.5) becomes

$$\overline{\sigma}((S, f)) = \overline{\sigma}(\{(x, f(x)) \mid x \in S\}) = \{(\sigma(x), f(x)) \mid x \in S\}.$$

In the case where G is the rotation group of the square acting on the vertices of the square, the example of acting on a coloring by ρ that we saw in (6.3) is an example of this kind of action. In the standard notation, when we act on a coloring by σ, the color in position i moves to position $\sigma(i)$.

Problem 285. Why does the action we have defined on colorings in Equation (6.5) take a coloring to a coloring?

Problem 286. Show that if G is a group of permutations of a set S, and f is a coloring function on S, then the equation

$$\overline{\overline{\sigma}}(\{(x, f(x)) \mid x \in S\}) = \{(\overline{\sigma}(x), f(x)) \mid x \in S\}$$

defines an action of G on the colorings (S, f) of S. (h)

6.2.2 Orbits

- **Problem 287.** Refer back to Problem 282 in answering the following questions.

    (a) What is the set of two element subsets that you get by computing $\overline{\sigma}(\{1,2\})$ for all σ in D_4?

    (b) What is the multiset of two-element subsets that you get by computing $\overline{\sigma}(\{1,2\})$ for all σ in D_4?

    (c) What is the set of two-element subsets you get by computing $\overline{\sigma}(\{1,3\})$ for all σ in D_4?

    (d) What is the multiset of two-element subsets that you get by computing $\overline{\sigma}(\{1,3\})$ for all σ in D_4?

    (e) Describe these two sets geometrically in terms of the square.

- **Problem 288.** This problem uses the notation for permutations in the dihedral group of the square introduced before Problem 259. What is the effect of a 180 degree rotation ρ^2 on the diagonals of a square? What is the effect of the flip $\varphi_{1|3}$ on the diagonals of a square? How many elements of D_4 send each diagonal to itself? How many elements of D_4 interchange the diagonals of a square?

In Problem 287 you saw that the action of the dihedral group D_4 on two element subsets of $\{1,2,3,4\}$ seems to split them into two sets, one with two elements and one with 4. We call these two sets the "orbits" of D_4 acting on the two elements subsets of $\{1,2,3,4\}$. More generally, the **orbit** of a permutation group G determined by an element x of a set S on which G acts is

$$\{\overline{\sigma}(x) | \sigma \in G\},$$

and is denoted by Gx. In Problem 287 it was possible to have $Gx = Gy$. In fact in that problem, $Gx = Gy$ for every y in Gx.

> **Problem 289.** Suppose a group acts on a set S. Could an element of S be in two different orbits? (Say why or why not.) (h)

Problem 289 almost completes the proof of the following theorem.

Theorem 6.2.3. *Suppose a group acts on a set S. The orbits of G form a partition of S.*

It is probably worth pointing out that this theorem tells us that the orbit Gx is also the orbit Gy for any element y of Gx.

Problem 290. Complete the proof of Theorem 6.2.3.

Notice that thinking in terms of orbits actually hides some information about the action of our group. When we computed the multiset of all results of acting on $\{1,2\}$ with the elements of D_4, we got an eight-element multiset containing each side twice. When we computed the multiset of all results of acting on $\{1,3\}$ with the elements of D_4, we got an eight-element multiset containing each diagonal of the square four times. These multisets remind us that we are acting on our two-element sets with an eight-element group. The **multiorbit** of G determined by an element x of S is the multiset

$$\{\overline{\sigma}(x) \mid \sigma \in G\},$$

and is denoted by Gx_{multi}.

When we used the quotient principle to count circular seating arrangements or necklaces, we partitioned up a set of lists of people or beads into blocks of equivalent lists. In the case of seating n people around a round table, what made two lists equivalent was, in retrospect, the action of the rotation group C_n. In the case of stringing n beads on a string to make a necklace, what made two lists equivalent was the action of the dihedral group. Thus the blocks of our partitions were orbits of the rotation group or the dihedral group, and we were counting the number of orbits of the group action. In Problem 45, we were not able to apply the quotient principle because we had blocks of different sizes. However, these blocks were still orbits of the action of the group D_4. And, even though the orbits have different sizes, we expect that each orbit corresponds naturally to a multiorbit and that the multiorbits all have the same size. Thus if we had a version of the quotient rule for a union of multisets, we could hope to use it to count the number of multiorbits.

Problem 291.

(a) Find the orbit and multiorbit of D_4 acting on the coloring

$$\{(1, R), (2, R), (3, B), (4, B)\},$$

or, in standard notation, $RRBB$ of the vertices of a square.

(b) How many group elements map the coloring $RRBB$ to itself? What is the multiplicity of $RRBB$ in its multiorbit?

(c) Find the orbit and multiorbit of D_4 acting on the coloring

$$\{(1, R), (2, B), (3, R), (4, B)\}.$$

(d) How many elements of the group send the coloring $RBRB$ to itself? What is the multiplicity of $RBRB$ in its orbit?

Problem 292.

(a) If G is a group, how is the set $\{\tau\sigma \mid \tau \in G\}$ related to G?

(b) Use this to show that y is in the multiorbit Gx_{multi} if and only if $Gx_{\text{multi}} = Gy_{\text{multi}}$.

Problem 292.b tells us that, when G acts on S, each element x of S is in one and only one multiorbit. Since each orbit is a subset of a multiorbit and each element x in S is in one and only one orbit, this also tells us there is a bijection between the orbits of G and the multiorbits of G, so that we have the same number of orbits as multiorbits.

When a group acts on a set, a group element is said to **fix** an element of $x \in S$ if $\overline{\sigma}(x) = x$. The set of all elements fixing an element x is denoted by $\text{Fix}(x)$.

Problem 293. Suppose a group G acts on a set S. What is special about the subset $\text{Fix}(x)$ for an element x of S?

• **Problem 294.** Suppose a group G acts on a set S. What is the relationship of the multiplicity of $x \in S$ in its multiorbit and the size of $\text{Fix}(x)$?

Problem 295. What can you say about relationships between the multiplicity of an element y in the multiorbit Gx_{multi} and the multiplicites of other elements? Try to use this to get a relationship between the size of an orbit of G and the size of G. (h)

We suggested earlier that a quotient principle for multisets might prove useful. The quotient principle came from the sum principle, and we do not have a sum principle for multisets. Such a principle would say that the size of a union of disjoint multisets is the sum of their sizes. We have not yet defined the union of multisets or disjoint multisets, because we haven't needed the ideas until now. We define the **union** of two multisets S and T to be the multiset in which the multiplicity of an element x is the maximum[5] of the multiplicity of x in S and its multiplicity in T. Similarly, the union of a family of multisets is defined by defining the multiplicity of an element x to be the maximum of its multiplicities in the members of the family. Two multisets are said to be **disjoint** if no element is a member of both, that is, if no element has multiplicity one or more in both. Since the size of a multiset is the sum of the multiplicities of its members, we immediately get the **sum principle for multisets**.

The size of a union of disjoint multisets is the sum of their sizes.

---

[5] We choose the maximum rather than the sum so that the union of sets is a special case of the union of multisets.

Taking the multisets all to have the same size, we get the **product principle** for multisets.

The union of a set of m disjoint multisets, each of size n has size mn.

The **quotient principle** for multisets then follows immediately.

If a p-element multiset is a union of q disjoint multisets, each of size r, then $q = p/r$.

- **Problem 296.** How does the size of the union of the set of multiorbits of a group G acting on a set S relate to the number of multiorbits and the size of G?

- **Problem 297.** How does the size of the union of the set of multiorbits of a group G acting on a set S relate to the numbers $|\text{Fix}(x)|$?

- **Problem 298.** In Problems 296 and 297 you computed the size of the union of the set of multiorbits of a group G acting on a set S in two different ways, getting two different expressions must be equal. Write the equation that says they are equal and solve for the number of multorbits, and therefore the number of orbits.

6.2.3 The Cauchy-Frobenius-Burnside Theorem

- **Problem 299.** In Problem 298 you stated and proved a theorem that expresses the number of orbits in terms of the number of group elements fixing each element of S. It is often easier to find the number of elements fixed by a given group element than to find the number of group elements fixing an element of S.

    (a) For this purpose, how does the sum $\sum_{x:\, x \in S} |\text{Fix}(x)|$ relate to the number of ordered pairs (σ, x) (with $\sigma \in G$ and $x \in S$) such that σ fixes x?

    (b) Let $\chi(\sigma)$ denote the number of elements of S fixed by σ. How can the number of ordered pairs (σ, x) (with $\sigma \in G$ and $x \in S$) such that σ fixes x be computed from $\chi(G)$? (It is ok to have a summation in your answer.)

    (c) What does this tell you about the number of orbits?

Problem 300. A second computation of the result of Problem 299 can be done as follows.

(a) Let $\widehat{\chi}(\sigma, x) = 1$ if $\sigma(x) = x$ and let $\widehat{\chi}(\sigma, x) = 0$ otherwise. Notice that $\widehat{\chi}$ is different from the χ in the previous problem, because it is a function of two variables. Use $\widehat{\chi}$ to convert the single summation in Problem 298 into a double summation over elements x of S and elements σ of G.

(b) Reverse the order of the previous summation in order to convert it into a single sum involving the function χ given by

$$\chi(\sigma) = \text{the number of elements of } S \text{ left fixed by } \sigma.$$

In Problem 299 you gave a formula for the number of orbits of a group G acting on a set X. This formula was first worked out by Cauchy in the case of the symmetric group, and then for more general groups by Frobenius. In his pioneering book on Group Theory, Burnside used this result as a lemma, and while he attributed the result to Cauchy and Frobenius in the first edition of his book, in later editions, he did not. Later on, other mathematicians who used his book named the result "Burnside's Lemma," which is the name by which it is still most commonly known. Let us agree to call this result the Cauchy-Frobenius-Burnside Theorem, or CFB Theorem for short in a compromise between historical accuracy and common usage.

⇒ **Problem 301.** In how many ways may we string four (identical) red, six (identical) blue, and seven (identical) green beads on a necklace? (h)

⇒ **Problem 302.** If we have an unlimited supply of identical red beads and identical blue beads, in how many ways may we string 17 of them on a necklace?

⇒ **Problem 303.** If we have five (identical) red, five (identical) blue, and five (identical) green beads, in how many ways may we string them on a necklace?

⇒ **Problem 304.** In how many ways may we paint the faces of a cube with six different colors, using all six?

Problem 305. In how many ways may we paint the faces of a cube with two colors of paint? What if both colors must be used? (h)

⇒ **Problem 306.** In how many ways may we color the edges of a (regular) $(2n + 1)$-gon free to move around in the *plane* (so it cannot be flipped) if we use red n times and blue $n + 1$ times? If this is a number you have seen before, identify it. (h)

⇒ ∗ **Problem 307.** In how many ways may we color the edges of a (regular) $(2n + 1)$-gon free to move in *three-dimensional space* so that n edges are colored red and $n + 1$ edges are colored blue. Your answer may depend on whether n is even or odd.

⇒ ∗ **Problem 308.** (Not unusually hard for someone who has worked on chromatic polynomials.) How many different proper colorings with four colors are there of the vertices of a graph which is cycle on five vertices? (If we get one coloring by rotating or flipping another one, they aren't really different.)

⇒ ∗ **Problem 309.** How many different proper colorings with four colors are there of the graph in Figure 6.2.4? Two graphs are the same if we can redraw one of the graphs, not changing the vertex set or edge set, so that it is identical to the other one. This is equivalent to permuting the vertices in some way so that when we apply the permutation to the endpoints of the edges to get a new edge set, the new edge set is equal to the old one. Such a permutation is called an **automorphism** of the graph. Thus two colorings are different if there is no automorphism of the graph that carries one to the other one.

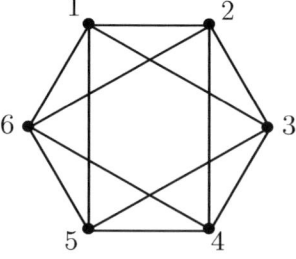

Figure 6.2.4: A graph on six vertices.

(h)

6.3 Pólya-Redfield Enumeration Theory

George Pólya and Robert Redfield independently developed a theory of generating functions that describe the action of a group G on functions from a set S to a set T when we know the action of G on S. Pólya's work on the subject is very accessible in its exposition, and so the subject has become popularly known as Pólya theory, though Pólya-Redfield theory would be a better name. In this section we develop the elements of this theory.

The idea of coloring a set S has many applications. For example, the set S might be the positions in a hydrocarbon molecule which are occupied by hydrogen, and the group could be the group of spatial symmetries of the molecule (that is, the group of permutations of the atoms of the molecule that move the molecule around so that in its final position the molecule cannot be distinguished from the original molecule). The colors could then be radicals (including hydrogen itself) that we could substitute for each hydrogen position in the molecule. Then the number of orbits of colorings is the number of chemically different compounds we could create by using these substitutions.[6]

In Figure 6.3.1 we show two different ways to substitute the OH radical for a hydrogen atom in the chemical diagram we gave for butane in Chapter 2. We have colored one vertex of degree 1 with the radical OH and the rest with the atom H. There are only two distinct ways to do this, as the OH must either connect to an "end" C or a "middle" C. This shows that there are two different forms, called isomers of the compound shown. This compound is known as butyl alcohol.

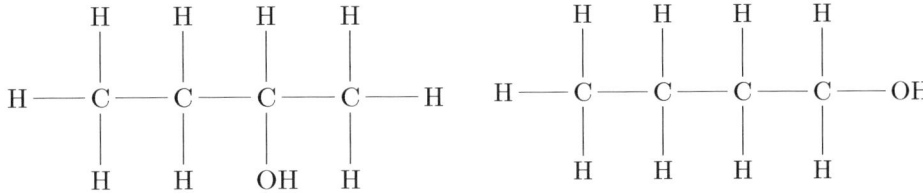

Figure 6.3.1: The two different isomers of butyl alcohol.

So think intuitively about some "figure" that has places to be colored. (Think of the faces of a cube, the beads on a necklace, circles at the vertices of an n-gon, etc.) How can we picture the coloring? If we number the places to be colored, say 1 to n, then a function from $[n]$ to the colors is exactly our coloring; if our colors are blue, green and red, then $BBGRRGBG$ describes a typical coloring of 8 such places. Unless the places are somehow "naturally" numbered, this idea of a coloring imposes structure that is not really there. Even if the structure is there, visualizing our colorings in this way doesn't "pull together" any common features

---

[6]There is a fascinating subtle issue of what makes two molecules different. For example, suppose we have a molecule in the form of a cube, with one atom at each vertex. If we interchange the top and bottom faces of the cube, each atom is still connected to exactly the same atoms as before. However we cannot achieve this permutation of the vertices by a member of the rotation group of the cube. It could well be that the two versions of the molecule interact with other molecules in different ways, in which case we would consider them chemically different. On the other hand if the two versions interact with other molecules in the same way, we would have no reason to consider them chemically different. This kind of symmetry is an example of what is called **chirality** in chemistry.

of different colorings; we are simply visualizing all possible functions. We have a group (think of it as symmetries of the figure you are imagining) that acts on the places. That group then acts in a natural way on the colorings of the places and we are interested in orbits of the colorings. Thus we want a picture that pulls together the common features of the colorings in an orbit. One way to pull together similarities of colorings would be to let the letters we are using as pictures of colors commute as we did with our pictures in Chapter 4; then our picture $BBGRRGBG$ becomes $B^3G^3R^2$, so our picture now records simply how many times we use each color. If you think about how we defined the action of a group on a set of functions, you will see that a group element won't change how many times each color is used; it simply moves colors to different places. Thus the picture we now have of a given coloring is an equally appropriate picture for each coloring in an orbit. One natural question for us to ask is "How many orbits have a given picture?"

Problem 310. Suppose we draw identical circles at the vertices of a regular hexagon. Suppose we color these circles with two colors, red and blue.

(a) In how many ways may we color the set $\{1, 2, 3, 4, 5, 6\}$ using the colors red and blue?

(b) These colorings are partitioned into orbits by the action of the rotation group on the hexagon. Using our standard notation, write down all these orbits and observe how many orbits have each picture, assuming the picture of a coloring is the product of commuting variables representing the colors.

(c) Using the picture function of the previous part, write down the picture enumerator for the orbits of colorings of the vertices of a hexagon under the action of the rotation group.

In Problem c we saw a picture enumerator for pictures of orbits of the action of a group on colorings. As above, we can ask how many orbits of the colorings have any given picture. We can think of a multivariable generating function in which the letters we use to picture individual colors are the variables, and the coefficient of a picture is the number of orbits with that picture. Such a generating function is an answer to our natural question, and so it is this sort of generating function we will seek. Since the CFB theorem was our primary tool for saying how many orbits we have, it makes sense to think about whether the CFB theorem has an analog in terms of pictures of orbits.

6.3.1 The Orbit-Fixed Point Theorem

- **Problem 311.** Suppose now we have a group G acting on a set and we have a picture function on that set with the additional feature that for each orbit of the group, all its elements have the same picture. In this circumstance we define the picture of an orbit or multiorbit to be the picture of any one of its members. The **orbit enumerator** $\text{Orb}(G, S)$ is the sum of the pictures of the orbits. (Note that this is the same as the sum of the pictures of the **multiorbits**.) The **fixed point enumerator** Fix(G, S) is the sum of the pictures of each of the fixed points of each of the elements of G. We are going to construct a generating function analog of the CFB theorem. The main idea of the proof of the CFB theorem was to try to compute in two different ways the number of elements (i.e. the sum of all the multiplicities of the elements) in the union of all the multiorbits of a group acting on a set. Suppose instead we try to compute the sum of all the pictures of all the elements in the union of the multiorbits of a group acting on a set. By thinking about how this sum relates to Orb(G, S) and Fix(G, S), find an analog of the CFB theorem that relates these two enumerators. State and prove this theorem.

We call the theorem of Problem 311 the **Orbit-Fixed Point Theorem**. In order to apply the Orbit-Fixed Point Theorem, we need some facts about picture enumerators.

- **Problem 312.** Suppose that P_1 and P_2 are picture functions on sets S_1 and S_2 in the sense of Section 4.1.2. Define P on $S_1 \times S_2$ by $P(x_1, x_2) = P_1(x_1)P_2(x_2)$. How are E_{P_1}, E_{P_1}, and E_P related? (You may have already done this problem in another context!)

- **Problem 313.** Suppose P_i is a picture function on a set S_i for $i = 1, \ldots, k$. We define the picture of a k-tuple $(x_1, x_2, \ldots, x_k)$ to be the product of the pictures of its elements, i.e.

$$\widehat{P}((x_1, x_2, \ldots, x_k)) = \prod_{i=1}^{k} P_i(x_i).$$

How does the picture enumerator $E_{\widehat{P}}$ of the set $S_1 \times S_2 \times \cdots \times S_k$ of all k-tuples with $x_i \in S$ relate to the picture enumerators of the sets S_i? In the special case that $S_i = S$ for all i and $P_i = P$ for all i, what is $E_{\widehat{P}}(S^k)$?

- **Problem 314.** Use the Orbit-Fixed Point Theorem to determine the Orbit Enumerator for the colorings, with two colors (red and blue), of six circles placed at the vertices of a hexagon which is free to move in the plane. Compare the coefficients of the resulting polynomial with the various orbits you found in Problem 310.

Problem 315. Find the generating function (in variables R, B) for colorings of the faces of a cube with two colors (red and blue). What does the generating function tell you about the number of ways to color the cube (up to spatial movement) with various combinations of the two colors.

6.3.2 The Pólya-Redfield Theorem

Pólya's (and Redfield's) famed enumeration theorem deals with situations such as those in Problems 314 and Problem 315 in which we want a generating function for the set of all functions from a set S to a set T on which a picture function is defined, and the picture of a function is the product of the pictures of its multiset of values. The point of the next series of problems is to analyze the solution to Problems 314 and Problem 315 in order to see what Pólya and Redfield saw (though they didn't see it in this notation or using this terminology).

- **Problem 316.** In Problem 314 we have four kinds of group elements: the identity (which fixes every coloring), the rotations through 60 or 300 degrees, the rotations through 120 and 240 degrees, and the rotation through 180 degrees. The fixed point enumerator for the rotation group acting on the functions is by definition the sum of the fixed point enumerators of colorings fixed by the identity, of colorings fixed by 60 or 300 degree rotations, of colorings fixed by 120 or 240 degree rotations, and of colorings fixed by the 180 degree rotation. Write down each of these enumerators (one for each kind of permutation) individually and factor each one (over the integers) as completely as you can.

- **Problem 317.** In Problem 315 we have five different kinds of group elements, and the fixed point enumerator is the sum of the fixed point enumerators of each of these kinds of group elements. For each kind of element, write down the fixed point enumerator for the elements of that kind. Factor the enumerators as completely as you can.

- **Problem 318.** In Problem 316, each "kind" of group element has a "kind" of cycle structure. For example, a rotation through 180 degrees has three cycles of size two. What kind of cycle structure does a rotation through 60 or 300 degrees have? What kind of cycle structure does a rotation through 120 or 240 degrees have? Discuss the relationship between the cycle structures and the factored enumerators of fixed points of the permutations in Problem 316.

Recall that we said that a group of permutations acts on a set if, for each member σ of G there is a bijection $\overline{\sigma}$ of S such that

$$\overline{\sigma \circ \varphi} = \overline{\sigma} \circ \overline{\varphi}$$

for every member σ and φ of G. Since $\overline{\sigma}$ is a bijection of S to itself, it is in fact a permutation of S. Thus $\overline{\sigma}$ has a cycle structure (that is, it is a product of disjoint cycles) as a permutation of S (in addition to whatever its cycle structure is in the original permutation group G).

- **Problem 319.** In Problem 317, each "kind" of group element has a "kind" of cycle structure in the action of the rotation group of the cube on the faces of the cube. For example, a rotation of the cube through 180 degrees around a vertical axis through the centers of the top and bottom faces has two cycles of size two and two cycles of size one. How many such rotations does the group have? What are the other "kinds" of group elements, and what are their cycle structures? Discuss the relationship between the cycle structure and the factored enumerator in Problem 317.

- **Problem 320.** The usual way of describing the Pólya-Redfield enumeration theorem involves the "cycle indicator" or "cycle index" of a group acting on a set. Suppose we have a group G acting on a finite set S. Since each group element σ gives us a permutation $\overline{\sigma}$ of S, as such it has a decomposition into disjoint cycles as a permutation of S. Suppose σ has c_1 cycles of size 1, c_2 cycles of size 2, ..., c_n cycles of size n. Then the **cycle monomial** of σ is

    $$z(\sigma) = z_1^{c_1} z_2^{c_2} \cdots z_n^{c_n}.$$

    The **cycle indicator** or **cycle index** of G acting on S is

    $$Z(G, S) = \frac{1}{|G|} \sum_{\sigma : \sigma \in G} z(\sigma).$$

    (a) What is the cycle index for the group D_6 acting on the vertices of a hexagon?

    (b) What is the cycle index for the group of rotations of the cube acting on the faces of the cube?

Problem 321. How can you compute the Orbit Enumerator of G acting on functions from S to a finite set T from the cycle index of G acting on S? (Use t, thought of as a variable, as the picture of an element t of T.) State and prove the relevant theorem! This is Pólya's and Redfield's famous enumeration theorem.

⇒ **Problem 322.** Suppose we make a necklace by stringing 12 pieces of brightly colored plastic tubing onto a string and fastening the ends of the string together. We have ample supplies blue, green, red, and yellow tubing available. Give a generating function in which the coefficient of $B^i G^j R^k Y^h$ is the number of necklaces we can make with i blues, j greens, k reds, and h yellows. How many terms would this generating function have if you expanded it in terms of powers of B, G, R, and Y? Does it make sense to do this expansion? How many of these necklaces have 3 blues, 3 greens, 2 reds, and 4 yellows?

Problem 323. What should we substitute for the variables representing colorings in the orbit enumerator of G acting on the set of colorings of S by a set T of colors in order to compute the total number of orbits of G acting on the set of colorings? What should we substitute into the variables in the cycle index of a group G acting on a set S in order to compute the total number of orbits of G acting on the colorings of S by a set T? Find the number of ways to color the faces of a cube with four colors.

⇒ **Problem 324.** We have red, green, and blue sticks all of the same length, with a dozen sticks of each color. We are going to make the skeleton of a cube by taking eight identical lumps of modeling clay and pushing three sticks into each lump so that the lumps become the vertices of the cube. (Clearly we won't need all the sticks!) In how many different ways could we make our cube? How many cubes have four edges of each color? How many have two red, four green, and six blue edges?

⇒ **Problem 325.** How many cubes can we make in Problem 324 if the lumps of modelling clay can be any of four colors?

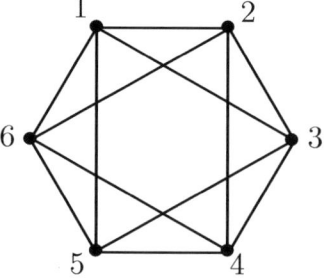

Figure 6.3.2: A possible computer network.

⇒ **Problem 326.** In Figure 6.3.2 we see a graph with six vertices. Suppose we have three different kinds of computers that can be placed at the six vertices of the graph to form a network. In how many different ways may the computers be placed? (Two graphs are not different if we can redraw one of the graphs so that it is identical to the other one.) This is equivalent to permuting the vertices in some way so that when we apply the permutation to the endpoints of the edges to get a new edge set, the new edge set is equal to the old one. Such a permutation is called an **automorphism** of the graph. Then two computer placements are the same if there is an automorphism of the graph that carries one to the other. (h)

⇒ **Problem 327.** Two simple graphs on the set $[n] = \{1, 2, \ldots, n\}$ with edge sets E and E' (which we think of a sets of two-element sets for this problem) are said to be **isomorphic** if there is a permutation σ of $[n]$ which, in its action of two-element sets, carries E to E'. We say two graphs are different if they are not isomorphic. Thus the number of different graphs is the number of orbits of the set of all two-element subsets of $[n]$ under the action of the group S_n. We can represent an edge set by its characteristic function (as in problem 33). That is we define

$$\chi_E(\{u,v\}) = \begin{cases} 1 & \text{if } \{u,v\} \in E \\ 0 & \text{otherwise.} \end{cases}$$

Thus we can think of the set of graphs as a set of colorings with colors 0 and 1 of the set of all two-element subsets of $[n]$. The number of different graphs with vertex set $[n]$ is thus the number of orbits of this set of colorings under the action of the symmetric group S_n on the set of two-element subsets of $[n]$. Use this to find the number of different graphs on five vertices. (h)

6.4 Supplementary Problems

1. Show that a function from S to T has an inverse (defined on T) if and only if it is a bijection.

2. How many elements are in the dihedral group D_3? The symmetric group S_3? What can you conclude about D_3 and S_3?

3. A tetrahedron is a thee dimensional geometric figure with four vertices, six edges, and four triangular faces. Suppose we start with a tetrahedron in space and consider the set of all permutations of the vertices of the tetrahedron that correspond to moving the tetrahedron in space and returning it to its original location, perhaps with the vertices in different places.
   (a) Explain why these permutations form a group.
   (b) What is the size of this group?

(c) Write down in two-row notation a permutation that is *not* in this group.

4. Find a three-element subgroup of the group S_3. Can you find a different three-element subgroup of S_3?

5. Prove true or demonstrate false with a counterexample: "In a permutation group, $(\sigma\varphi)^n = \sigma^n\varphi^n$."

6. If a group G acts on a set S, and if $\sigma(x) = y$, is there anything interesting we can say about the subgroups $\text{Fix}(x)$ and $\text{Fix}(y)$?

7.
   (a) If a group G acts on a set S, does $\overline{\sigma}(f) = f \circ \sigma$ define a group action on the functions from S to a set T? Why or why not?

   (b) If a group G acts on a set S, does $\sigma(f) = f \circ \sigma^{-1}$ define a group action on the functions from S to a set T? Why or why not?

   (c) Is either of the possible group actions essentially the same as the action we described on colorings of a set, or is that an entirely different action?

8. Find the number of ways to color the faces of a tetrahedron with two colors.

9. Find the number of ways to color the faces of a tetrahedron with four colors so that each color is used.

10. Find the cycle index of the group of spatial symmetries of the tetrahedron acting on the vertices. Find the cycle index for the same group acting on the faces.

11. Find the generating function for the number of ways to color the faces of the tetrahedron with red, blue, green and yellow.

$\Rightarrow$ **12.** Find the generating function for the number of ways to color the faces of a cube with four colors so that all four colors are used.

$\Rightarrow$ **13.** How many different graphs are there on six vertices with seven edges?

$\Rightarrow$ **14.** Show that if H is a subgroup of the group G, then H acts on G by $\sigma(\tau) = \sigma \circ \tau$ for all σ in H and τ in G. What is the size of an orbit of this action? How does the size of a subgroup of a group relate to the size of the group?

Appendix A

Relations

A.1 Relations as sets of Ordered Pairs

A.1.1 The relation of a function

> **Problem 328.** Consider the functions from $S = \{-2, -1, 0, 1, 2\}$ to $T = \{1, 2, 3, 4, 5\}$ defined by $f(x) = x + 3$, and $g(x) = x^5 - 5x^3 + 5x + 3$. Write down the set of ordered pairs $(x, f(x))$ for $x \in S$ and the set of ordered pairs $(x, g(x))$ for $x \in S$. Are the two functions the same or different?

Problem 328 points out how two functions which appear to be different are actually the same on some domain of interest to us. Most of the time when we are thinking about functions it is fine to think of a function casually as a relationship between two sets. In Problem 328 the set of ordered pairs you wrote down for each function is called the **relation** of the function. When we want to distinguish between the casual and the careful in talking about relationships, our casual term will be "relationship" and our careful term will be "relation." So *relation* is a technical word in mathematics, and as such it has a technical definition. A **relation** from a set S to a set T is a set of ordered pairs whose first elements are in S and whose second elements are in T. Another way to say this is that a **relation** from S to T is a subset of $S \times T$.

A typical way to define a **function** f from a set S, called the **domain** of the function, to a set T, called the **range**, is that f is a relationship between S to T that relates one and only one member of T to each element of X. We use $f(x)$ to stand for the element of T that is related to the element x of S. If we wanted to make our definition more precise, we could substitute the word "relation" for the word "relationship" and we would have a more precise definition. For our purposes, you can choose whichever definition you prefer. However, in any case, there is a relation associated with each function. As we said above, the relation of a function $f : S \to T$ (which is the standard shorthand for "f is a function from S to T" and is usually read as f **maps** S to T) is the set of all ordered pairs $(x, f(x))$ such that x is in S.

Problem 329. Here are some questions that will help you get used to the formal idea of a relation and the related formal idea of a function. S will stand for a set of size s and T will stand for a set of size t.

  (a) What is the size of the largest relation from S to T?

  (b) What is the size of the smallest relation from S to T?

  (c) The relation of a function $f : S \to T$ is the set of all ordered pairs $(x, f(x))$ with $x \in S$. What is the size of the relation of a function from S to T? That is, how many ordered pairs are in the relation of a function from S to T? (h)

  (d) We say f is a **one-to-one** function or **injection** from S to T if each member of S is related to a *different* element of T. How many different elements must appear as second elements of the ordered pairs in the relation of a one-to-one function (injection) from S to T?

  (e) A function $f : S \to T$ is called an **onto function** or **surjection** if each element of T is $f(x)$ for some $x \in S$. What is the minimum size that S can have if there is a surjection from S to T?

Problem 330. When f is a function from S to T, the sets S and T play a big role in determining whether a function is one-to-one or onto (as defined in Problem 329). For the remainder of this problem, let S and T stand for the set of nonnegative real numbers.

  (a) If $f : S \to T$ is given by $f(x) = x^2$, is f one-to-one? Is f onto?

  (b) Now assume S' is the set of all real numbers and $g : S' \to T$ is given by $g(x) = x^2$. Is g one-to-one? Is g onto?

  (c) Assume that T' is the set of all real numbers and $h : S \to T'$ is given by $h(x) = x^2$. Is h one-to-one? Is h onto?

  (d) And if the function $j : S' \to T'$ is given by $j(x) = x^2$, is j one-to-one? Is j onto?

Problem 331. If $f : S \to T$ is a function, we say that f **maps** x to y as another way to say that $f(x) = y$. Suppose $S = T = \{1, 2, 3\}$. Give a function from S to T that is not onto. Notice that two different members of S have mapped to the same element of T. Thus when we say that f associates one and only one element of T to each element of S, it is quite possible that the one and only one element $f(1)$ that f maps 1 to is exactly the same as the one and only one element $f(2)$ that f maps 2 to.

A.1.2 Directed graphs

We visualize numerical functions like $f(x) = x^2$ with their graphs in Cartesian coordinate systems. We will call these kinds of graphs *coordinate graphs* to distinguish them from other kinds of graphs used to visualize relations that are non-numerical.

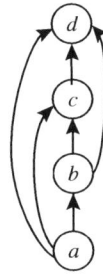

Figure A.1.1: The alphabet digraph.

In Figure A.1.1 we illustrate another kind of graph, a "directed graph" or "digraph" of the "comes before in alphabetical order" relation on the letters a, b, c, and d. To draw a **directed graph** of a relation on a set S, we draw a circle (or dot, if we prefer), which we call a **vertex**, for each element of the set, we usually label the vertex with the set element it corresponds to, and we draw an arrow from the vertex for a to that for b if a is related to b, that is, if the ordered pair (a, b) is in our relation. We call such an arrow an **edge** or a **directed edge**. We draw the arrow from a to b, for example, because a comes before b in alphabetical order. We try to choose the locations where we draw our vertices so that the arrows capture what we are trying to illustrate as well as possible. Sometimes this entails redrawing our directed graph several times until we think the arrows capture the relationship well.

We also draw digraphs for relations from a set S to a set T; we simply draw vertices for the elements of S (usually in a row) and vertices for the elements of T (usually in a parallel row) draw an arrow from x in S to y in T if x is related to y. Notice that instead of referring to the vertex representing x, we simply referred to x. This is a common shorthand. Here are some exercises just to practice drawing digraphs.

> **Problem 332.** Draw the digraph of the "is a proper subset of" relation on the set of subsets of a two element set. How many arrows would you have had to draw if this problem asked you to draw the digraph for the subsets of a three-element set? (h)

We also draw digraphs for relations from finite set S to a finite set T; we simply draw vertices for the elements of S (usually in a row) and vertices for the elements of T (usually in a parallel row) and draw an arrow from x in S to y in T if x is related to y. Notice that instead of referring to the vertex representing x, we simply referred to x. This is a common shorthand.

Problem 333. Draw the digraph of the relation from the set {A, M, P, S} to the set {Sam, Mary, Pat, Ann, Polly, Sarah} given by "is the first letter of."

Problem 334. Draw the digraph of the relation from the set {Sam, Mary, Pat, Ann, Polly, Sarah} to the set {A, M, P, S} given by "has as its first letter."

Problem 335. Draw the digraph of the relation on the set {Sam, Mary, Pat, Ann, Polly, Sarah} given by "has the same first letter as."

A.1.3 Digraphs of Functions

Problem 336. When we draw the digraph of a function f, we draw an arrow *from* the vertex representing x *to* the vertex representing $f(x)$. One of the relations you considered in Problems 333 and Problem 334 is the relation of a function.

(a) Which relation is the relation of a function?

(b) How does the digraph help you visualize that one relation is a function and the other is not?

Problem 337. Digraphs of functions help us to visualize whether or not they are onto or one-to-one. For example, let both S and T be the set $\{-2, -1, 0, 1, 2\}$ and let S' and T' be the set $\{0, 1, 2\}$. Let $f(x) = 2 - |x|$.

(a) Draw the digraph of the function f assuming its domain is S and its range is T. Use the digraph to explain why or why not this function maps S onto T.

(b) Use the digraph of the previous part to explain whether or not the function is one-to one.

(c) Draw the digraph of the function f assuming its domain is S and its range is T'. Use the digraph to explain whether or not the function is onto.

(d) Use the digraph of the previous part to explain whether or not the function is one-to-one.

(e) Draw the digraph of the function f assuming its domain is S' and its range is T'. Use the digraph to explain whether the function is onto.

(f) Use the digraph of the previous part to explain whether the function is one-to-one.

(g) Suppose the function f has domain S' and range T. Draw the digraph of f and use it to explain whether f is onto.

(h) Use the digraph of the previous part to explain whether f is one-to-one.

A one-to-one function from a set X onto a set Y is frequently called a **bijection**, especially in combinatorics. Your work in Problem 337 should show you that a digraph is the digraph of a bijection from X to Y

- if the vertices of the digraph represent the elements of X and Y,

- if each vertex representing an element of X has one and only one arrow leaving it, and

- each vertex representing an element of Y has one and only one arrow entering it.

Problem 338. If we reverse all the arrows in the digraph of a bijection f, we get the digraph of another function g. Is g a bijection? What is $f(g(x))$? What is $g(f(x))$?

If f is a function from S to T, if g is a function from T to S, and if $f(g(x)) = x$ for each x in T and $g(f(x)) = x$ for each x in S, then we say that g is an inverse of f (and f is an inverse of g).

More generally, if f is a function from a set R to a set S, and g is a function from S to T, then we define a new function $f \circ g$, called the **composition** of f and g, by $f \circ g(x) = f(g(x))$. Composition of functions is a particularly important operatio in subjects such as calculus, where we represent a function like $h(x) = \sqrt{x^2 + 1}$ as the composition of the square root function and the square and add one function in order to use the chain rule to take the derivative of h.

The function ι (the Greek letter iota is pronounced eye-oh-ta) from a set S to itself, given by the rule $\iota(x) = x$ for every x in S, is called the **identity function** on S. If f is a function from S to T and g is a function from T to S such that $g(f(x)) = x$ for every x in S, we can express this by saying that $g \circ f = \iota$, where ι is the identity function of S. Saying that $f(g(x)) = x$ is the same as saying that $f \circ g = \iota$, where ι stands for the identity function on T. We use the same letter for the identity function on two different sets when we can use context to tell us on which set the identity function is being defined.

Problem 339. If f is a function from S to T and g is a function from T to S such that $g(f(x)) = x$, how can we tell from context that $g \circ f$ is the identity function on S and not the identity function on T? (h)

Problem 340. Explain why a function that has an inverse must be a bijection.

Problem 341. Is it true that the inverse of a bijection is a bijection?

Problem 342. If g and h are inverse of f, then what can we say about g and h?

Problem 343. Explain why a bijection must have an inverse.

Since a function with an inverse has exactly one inverse g, we call g *the* inverse of f. From now on, when f has an inverse, we shall denote its inverse by f^{-1}. Thus $f(f^{-1}(x)) = x$ and $f^{-1}(f(x)) = x$. Equivalenetly $f \circ f^{-1} = \iota$ and $f^{-1} \circ f = \iota$.

A.2 Equivalence relations

So far we've used relations primarily to talk about functions. There is another kind of relation, called an equivalence relation, that comes up in the counting problems with which we began. In Problem 8 with three distinct flavors, it was probably tempting to say there are 12 flavors for the first pint, 11 for the second, and 10 for the third, so there are $12 \cdot 11 \cdot 10$ ways to choose the pints of ice cream. However, once the pints have been chosen, bought, and put into a bag, there is no way to tell which is first, which is second and which is third. What we just counted is lists of three distinct flavors—one to one functions from the set $\{1, 2, 3\}$ in to the set of ice cream flavors. Two of those lists become equivalent once the ice cream purchase is made if they list the same ice cream. In other words, two of those lists become equivalent (are related) if they list same subset of the set of ice cream flavors. To visualize this relation with a digraph, we would need one vertex for each of the $12 \cdot 11 \cdot 10$ lists. Even with five flavors of ice cream, we would need one vertex for each of $5 \cdot 4 \cdot 3 = 60$ lists. So for now we will work with the easier to draw question of choosing three pints of ice cream of different flavors from four flavors of ice cream.

Problem 344. Suppose we have four flavors of ice cream, V(anilla), C(hocolate), S(trawberry) and P(each). Draw the directed graph whose vertices consist of all lists of three distinct flavors of the ice cream, and whose edges connect two lists if they list the same three flavors. This graph makes it pretty clear in how many ways we may choose 3 flavors out of four. How many is it?

⇒ **Problem 345.** Now suppose again we are choosing three distinct flavors of ice cream out of four, but instead of putting scoops in a cone or choosing pints, we are going to have the three scoops arranged symmetrically in a circular dish. Similarly to choosing three pints, we can describe a selection of ice cream in terms of which one goes in the dish first, which one goes in second (say to the right of the first), and which one goes in third (say to the right of the second scoop, which makes it to the left of the first scoop). But again, two of these lists will sometimes be equivalent. Once they are in the dish, we can't tell which one went in first. However, there is a subtle difference between putting each flavor in its own small dish and putting all three flavors in a circle in a larger dish. Think about what makes the lists of flavors equivalent, and draw the directed graph whose vertices consist of all lists of three of the flavors of ice cream and whose edges connect two lists that we cannot tell the difference between as dishes of ice cream. How many dishes of ice cream can we distinguish from one another? (h)

Problem 346. Draw the digraph for Problem 38 in the special case where we have four people sitting around the table.

In Problems 344, 345, and 346 (as well as Problems 34, 38, and 39) we can begin with a set of lists, and say when two lists are equivalent as representations of the objects we are trying to count. In particular, in Problems 344, 345, and 346 you drew the directed graph for this relation of equivalence. Technically, you should have had an arrow from each vertex (list) to itself. This is what we mean when we say a relation is **reflexive**. Whenever you had an arrow from one vertex to a second, you had an arrow back to the first. This is what we mean when we say a relation is **symmetric**.

When people sit around a round table, each list is equivalent to itself: if List1 and List 2 are identical, then everyone has the same person to the right in both lists (including the first person in the list being to the right of the last person). To see the symmetric property of the equivalence of seating arrangements, if List1 and List2 are different, but everyone has the same person to the right when they sit according to List2 as when they sit according to List1, then everybody better have the same person to the right when they sit according to List1 as when they sit according to List2.

In Problems 344, 345 and 346 there is another property of those relations you may have noticed from the directed graph. Whenever you had an arrow from L_1 to L_2 and an arrow from L_2 to L_3, then there was an arrow from L_1 to L_3. This is what we mean when we say a relation is **transitive**. You also undoubtedly noticed how the directed graph divides up into clumps of mutually connected vertices. This is what equivalence relations are all about. Let's be a bit more precise in our description of what it means for a relation to be reflexive, symmetric or transitive.

- If R is a relation on a set X, we say R is **reflexive** if $(x, x) \in R$ for every $x \in X$.

- If R is a relation on a set X, we say R is **symmetric** if (x, y) is in R whenever (y, x) is in R.

- If R is a relation on a set X, we say R is **transitive** if whenever (x, y) is in R and (y, z) is in R, then (x, z) is in R as well.

Each of the relations of equivalence you worked with in Problems 344, 345 and 346 had these three properties. Can you visualize the same three properties in the relations of equivalence that you would use in Problems 34, 38, and 39? We call a relation an **equivalence relation** if it is reflexive, symmetric and transitive.

After some more examples, we will see how to show that equivalence relations have the kind of clumping property you saw in the directed graphs. In our first example, using the notation $(a, b) \in R$ to say that a is related to B is going to get in the way. It is really more common to write aRb to mean that a is related to b. For example, if our relation is the less than relation on $\{1, 2, 3\}$, you are much more likely to use $x < y$ than you are $(x, y) \in <$, aren't you? The reflexive law then says xRx for every x in X, the symmetric law says that if xRy, then yRx, and the transitive law says that if xRy and yRz, then xRz.

Problem 347. For the necklace problem, Problem 43, our lists are lists of beads. What makes two lists equivalent for the purpose of describing a necklace? Verify explicitly that this relationship of equivalence is reflexive, symmetric, and transitive.

Problem 348. Which of the reflexive, symmetric and transitive properties does the $<$ relation on the integers have?

Problem 349. A relation R on the set of ordered pairs of positive integers that you learned about in grade school in another notation is the relation that says (m, n) is related to (h, k) if $mk = hn$. Show that this relation is an equivalence relation. In what context did you learn about this relation in grade school? (h)

Problem 350. Another relation that you may have learned about in school, perhaps in the guise of "clock arithmetic," is the relation of equivalence modulo n. For integers (positive, negative, or zero) a and b, we write $a \equiv b \pmod{n}$ to mean that $a - b$ is an integer multiple of n, and in this case, we say that a is **congruent to** b **modulo** n and write $a \equiv b \pmod{n}$.. Show that the relation of congruence modulo n is an equivalence relation.

Problem 351. Define a relation on the set of all lists of n distinct integers chosen from $\{1, 2, \ldots, n\}$, by saying two lists are related if they have the same elements (though perhaps in a different order) in the first k places, and the same elements (though perhaps in a different order) in the last $n - k$ places. Show this relation is an equivalence relation.

Problem 352. Suppose that R is an equivalence relation on a set X and for each $x \in X$, let $C_x = \{y | y \in X \text{ and } yRx\}$. If C_x and C_z have an element y in common, what can you conclude about C_x and C_z (besides the fact that they have an element in common!)? Be explicit about what property(ies) of equivalence relations justify your answer. Why is every element of X in some set C_x? Be explicit about what property(ies) of equivalence relations you are using to answer this question. Notice that we might simultaneously denote a set by C_x and C_y. Explain why the union of the sets C_x is X. Explain why two distinct sets C_x and C_z are disjoint. What do these sets have to do with the "clumping" you saw in the digraph of Problem 344 and Problem 345?

In Problem 352 the sets C_x are called **equivalence classes** of the equivalence relation R. You have just proved that if R is an equivalence relation of the set X, then each element of X is in exactly one equivalence class of R. Recall that a **partition** of a set X is a set of disjoint sets whose union is X. For example, $\{1, 3\}$, $\{2, 4, 6\}$, $\{5\}$ is a partition of the set $\{1, 2, 3, 4, 5, 6\}$. Thus another way to describe what you proved in Problem 352 is the following:

Theorem A.2.1. *If R is an equivalence relation on X, then the set of equivalence classes of R is a partition of X.*

Since a partition of S is a set of subsets of S, it is common to call the subsets into which we partition S the **blocks** of the partition so that we don't find ourselves in the uncomfortable position of referring to a set and not being sure whether it is the set being partitioned or one of the blocks of the partition.

Problem 353. In each of Problems 38, Problem 39, Problem 43, Problem 344, and Problem 345, what does an equivalence class correspond to? (Five answers are expected here.) (h)

Problem 354. Given the partition $\{1, 3\}$, $\{2, 4, 6\}$, $\{5\}$ of the set $\{1, 2, 3, 4, 5, 6\}$, define two elements of $\{1, 2, 3, 4, 5, 6\}$ to be related if they are in the same part of the partition. That is, define 1 to be related to 3 (and 1 and 3 each related to itself), define 2 and 4, 2 and 6, and 4 and 6 to be related (and each of 2, 4, and 6 to be related to itself), and define 5 to be related to itself. Show that this relation is an equivalence relation.

Problem 355. Suppose $P = \{S_1, S_2, S_3, \ldots, S_k\}$ is a partition of S. Define two elements of S to be related if they are in the same set S_i, and otherwise not to be related. Show that this relation is an equivalence relation. Show that the equivalence classes of the equivalence relation are the sets S_i.

In Problem 353 you just proved that each partition of a set gives rise to an equivalence relation whose classes are just the parts of the partition. Thus in Problem 352 and Problem 353 you proved the following Theorem.

Theorem A.2.2. *A relation R is an equivalence relation on a set S if and only if S may be partitioned into sets $S_1, S_2, \ldots, S_n$ in such a way that x and y are related by R if and only if they are in the same block S_i of the partition.*

In Problems 344, Problem 345, Problem 38 and Problem 43 what we were doing in each case was counting equivalence classes of an equivalence relation. There was a special structure to the problems that made this somewhat easier to do. For example, in Problem 344, we had $4 \cdot 3 \cdot 2 = 24$ lists of three distinct flavors chosen from V, C, S, and P. Each list was equivalent to $3 \cdot 2 \cdot 1 = 3! = 6$ lists, including itself, from the point of view of serving 3 small dishes of ice cream. The order in which we selected the three flavors was unimportant. Thus the set of all $4 \cdot 3 \cdot 2$ lists was a union of some number n of equivalence classes, each of size 6. By the product principle, if we have a union of n disjoint sets, each of size 6, the union has $6n$ elements. But we already knew that the union was the set of all 24 lists of three distinct letters chosen from our four letters. Thus we have $6n = 24$, or $n = 4$ equivalence classes.

In Problem 345 there is a subtle change. In the language we adopted for seating people around a round table, if we choose the flavors V, C, and S, and arrange them in the dish with C to the right of V and S to the right of C, then the scoops are in different relative positions than if we arrange them instead with S to the right of V and C to the right of S. Thus the order in which the scoops go into the dish is somewhat important—somewhat, because putting in V first, then C to its right and S to its right is the same as putting in S first, then V to its right and C to its right. In this case, each list of three flavors is equivalent to only three lists, including itself, and so if there are n equivalence classes, we have $3n = 24$, so there are $24/3 = 8$ equivalence classes.

Problem 356. If we have an equivalence relation that divides a set with k elements up into equivalence classes each of size m, what is the number n of equivalence classes? Explain why.

Problem 357. In Problem 347, what is the number of equivalence classes? Explain in words the relationship between this problem and the Problem 39.

Problem 358. Describe explicitly what makes two lists of beads equivalent in Problem 43 and how Problem 356 can be used to compute the number of different necklaces.

Problem 359. What are the equivalence classes (write them out as sets of lists) in Problem 45, and why can't we use Problem 356 to compute the number of equivalence classes?

In Problem 356 you proved our next theorem. In Chapter 1 (Problem 42) we discovered and stated this theorem in the context of partitions and called it the **Quotient Principle**

Theorem A.2.3. *If an equivalence relation on a set S size k has n equivalence classes each of size m, then the number of equivalence classes is k/m.*

Appendix B

Mathematical Induction

B.1 The Principle of Mathematical Induction

B.1.1 The ideas behind mathematical induction

There is a variant of one of the bijections we used to prove the Pascal Equation that comes up in counting the subsets of a set. In the next problem it will help us compute the total number of subsets of a set, regardless of their size. Our main goal in this problem, however, is to introduce some ideas that will lead us to one of the most powerful proof techniques in combinatorics (and many other branches of mathematics), the principle of mathematical induction.

Problem 360.

(a) Write down a list of the subsets of $\{1,2\}$. Don't forget the empty set! Group the sets containing containing 2 separately from the others.

(b) Write down a list of the subsets of $\{1,2,3\}$. Group the sets containing 3 separately from the others.

(c) Look for a natural way to match up the subsets containing 2 in Part a with those not containing 2. Look for a way to match up the subsets containing 3 in Part b containing 3 with those not containing 3.

(d) On the basis of the previous part, you should be able to find a bijection between the collection of subsets of $\{1, 2, \ldots, n\}$ containing n and those not containing n. (If you are having difficulty figuring out the bijection, try rethinking Parts a and b, perhaps by doing a similar exercise with the set $\{1, 2, 3, 4\}$.) Describe the bijection (unless you are very familiar with the notation of sets, it is probably easier to describe to describe the function in words rather than symbols) and explain why it is a bijection. Explain why the number of subsets of $\{1, 2, \ldots, n\}$ containing n equals the number of subsets of $\{1, 2, \ldots, n-1\}$.

(e) Parts a and b suggest strongly that the number of subsets of a n-element set is 2^n. In particular, the empty set has 2^0 subsets, a one-element set has 2^1 subsets, itself and the empty set, and in Parts a and b we saw that two-element and three-element sets have 2^2 and 2^3 subsets respectively. So there are certainly some values of n for which an n-element set has 2^n subsets. One way to prove that an n-element set has 2^n subsets for all values of n is to argue by contradiction. For this purpose, suppose there is a nonnegative integer n such that an n-element set doesn't have exactly 2^n subsets. In that case there may be more than one such n. Choose k to be the smallest such n. Notice that $k-1$ is still a positive integer, because k can't be 0, 1, 2, or 3. Since k was the smallest value of n we could choose to make the statement "An n-element set has 2^n subsets" false, what do you know about the number of subsets of a $(k-1)$-element set? What do you know about the number of subsets of the k-element set $\{1, 2, \ldots, k\}$ that don't contain k? What do you know about the number of subsets of $\{1, 2, \ldots, k\}$ that do contain k? What does the sum principle tell you about the number of subsets of $\{1, 2, \ldots, k\}$? Notice that this contradicts the way in which we chose k, and the only assumption that went into our choice of k was that "there is a nonnegative integer n such that an n-element set doesn't have exactly 2^n subsets." Since this assumption has led us to a contradiction, it must be false. What can you now conclude about the statement "for every nonnegative integer n, an n-element set has exactly 2^n subsets?"

Problem 361. The expression

$$1 + 3 + 5 + \cdots + 2n - 1$$

is the sum of the first n odd integers. Experiment a bit with the sum for the first few positive integers and guess its value in terms of n. Now apply the technique of Problem 360 to prove that you are right. (h)

In Problems 360 and 361 our proofs had several distinct elements. We had a statement involving an integer n. We knew the statement was true for the first few nonnegative integers in Problem 360 and for the first few positive integers in Problem 361. We wanted to prove that the statement was true for all nonnegative integers in Problem 360 and for all positive integers in Problem 361. In both cases we used the method of proof by contradiction; for that purpose we assumed that there was a value of n for which our formula wasn't true. We then chose k to be the smallest value of n for which our formula wasn't true. This meant that when n was $k-1$, our formula was true, (or else that $k-1$ wasn't a nonnegative integer in Problem 360 or that $k-1$ wasn't a positive integer in Problem 361). What we did next was the crux of the proof. We showed that the truth of our statement for $n = k-1$ implied the truth of our statement for $n = k$. This gave us a contradiction to the assumption that there was an n that made the statement false. In fact, we will see that we can bypass entirely the use of proof by contradiction. We used it to help you discover the central ideas of the technique of proof by mathematical induction.

The central core of mathematical induction is the proof that the truth of a statement about the integer n for $n = k - 1$ implies the truth of the statement for $n = k$. For example, once we know that a set of size 0 has 2^0 subsets, if we have proved our implication, we can then conclude that a set of size 1 has 2^1 subsets, from which we can conclude that a set of size 2 has 2^2 subsets, from which we can conclude that a set of size 3 has 2^3 subsets, and so on up to a set of size n having 2^n subsets for any nonnegative integer n we choose. In other words, although it was the idea of proof by contradiction that led us to think about such an implication, we can now do without the contradiction at all. What we need to prove a statement about n by this method is a place to start, that is a value b of n for which we know the statement to be true, and then a proof that the truth of our statement for $n = k - 1$ implies the truth of the statement for $n = k$ whenever $k > b$.

B.1.2   Mathematical induction

The **principle of mathematical induction** states that

> In order to prove a statement about an integer n, if we can
>
> 1. Prove the statement when $n = b$, for some fixed integer b
> 2. Show that the truth of the statement for $n = k-1$ implies the truth of the statement for $n = k$ whenever $k > b$,
>
> then we can conclude the statement is true for all integers $n \geq b$.

As an example, let us return to Problem 360. The statement we wish to prove is the statement that "A set of size n has 2^n subsets."

> Our statement is true when $n = 0$, because a set of size 0 is the empty set and the empty set has $1 = 2^0$ subsets. (This step of our proof is called a **base step**.) Now suppose that $k > 0$ and every set with $k - 1$ elements has 2^{k-1} subsets. Suppose $S = \{a_1, a_2, \ldots a_k\}$ is a set with k elements. We partition the subsets of S into two blocks. Block B_1 consists of the subsets that do not contain a_n and block B_2 consists of the subsets that do contain a_n. Each set in B_1 is a subset of $\{a_1, a_2, \ldots a_{k-1}\}$, and each subset of $\{a_1, a_2, \ldots a_{k-1}\}$ is in B_1. Thus B_1 is the set of all subsets of $\{a_1, a_2, \ldots a_{k-1}\}$. Therefore by our assumption in the first sentence of this paragraph, the size of B_1 is 2^{k-1}. Consider the function from B_2 to B_1 which takes a subset of S including a_n and removes a_n from it. This function is defined on B_2, because every set in B_2 contains a_n. This function is onto, because if T is a set in B_1, then $T \cup \{a_k\}$ is a set in B_2 which the function sends to T. This function is one-to-one because if V and W are two different sets in B_2, then removing a_k from them gives two different sets in B_1. Thus we have a bijection between B_1 and B_2, so B_1 and B_2 have the same size. Therefore by the sum principle the size of $B_1 \cup B_2$ is $2^{k-1} + 2^{k-1} = 2^k$. Therefore S has 2^k subsets. This

shows that if a set of size $k-1$ has 2^{k-1} subsets, then a set of size k has 2^k subsets. Therefore by the principle of mathematical induction, a set of size n has 2^n subsets for every nonnegative integer n.

The first sentence of the last paragraph is called the **inductive hypothesis**. In an inductive proof we always make an inductive hypothesis as part of proving that the truth of our statement when $n = k - 1$ implies the truth of our statement when $n = k$. The last paragraph itself is called the **inductive step** of our proof. In an inductive step we derive the statement for $n = k$ from the statement for $n = k - 1$, thus proving that the truth of our statement when $n = k - 1$ implies the truth of our statement when $n = k$. The last sentence in the last paragraph is called the **inductive conclusion**. All inductive proofs should have a base step, an inductive hypothesis, an inductive step, and an inductive conclusion.

There are a couple details worth noticing. First, in this problem, our base step was the case $n = 0$, or in other words, we had $b = 0$. However, in other proofs, b could be any integer, positive, negative, or 0. Second, our proof that the truth of our statement for $n = k - 1$ implies the truth of our statement for $n = k$ required that k be at least 1, so that there would be an element a_k we could take away in order to describe our bijection. However, condition (2) of the principle of mathematical induction only requires that we be able to prove the implication for $k > 0$, so we were allowed to assume $k > 0$.

Problem 362. Use mathematical induction to prove your formula from Problem 361.

B.1.3 Proving algebraic statements by induction

Problem 363. Use mathematical induction to prove the well-known formula that for all positive integers n,

$$1 + 2 + \cdots + n = \frac{n(n+1)}{2}.$$

Problem 364. Experiment with various values of n in the sum

$$\frac{1}{1 \cdot 2} + \frac{1}{2 \cdot 3} + \frac{1}{3 \cdot 4} + \cdots + \frac{1}{n \cdot (n+1)} = \sum_{i=1}^{n} \frac{1}{i \cdot (i+1)}.$$

Guess a formula for this sum and prove your guess is correct by induction.

Problem 365. For large values of n, which is larger, n^2 or 2^n? Use mathematical induction to prove that you are correct. (h)

Problem 366. What is wrong with the following attempt at an inductive proof that all integers in any consecutive set of n integers are equal for every positive integer n? For an arbitrary integer i, all integers from i to i are equal, so our statement is true when $n = 1$. Now suppose $k > 1$ and all integers in any consecutive set of $k - 1$ integers are equal. Let S be a set of k consecutive integers. By the inductive hypothesis, the first $k - 1$ elements of S are equal and the last $k - 1$ elements of S are equal. Therefore all the elements in the set S are equal. Thus by the principle of mathematical induction, for every positive n, every n consecutive integers are equal. (h)

B.1.4 Strong Induction

One way of looking at the principle of mathematical induction is that it tells us that if we know the "first" case of a theorem and we can derive each other case of the theorem from a smaller case, then the theorem is true in all cases. However the particular way in which we stated the theorem is rather restrictive in that it requires us to derive each case from the immediately preceding case. This restriction is not necessary, and removing it leads us to a more general statement of the principal of mathematical induction which people often call the **strong principle of mathematical induction**. It states:

In order to prove a statement about an integer n if we can

1. prove our statement when $n = b$ and
2. prove that the statements we get with $n = b, n = b+1, \ldots n = k-1$ imply the statement with $n = k$,

then our statement is true for all integers $n \geq b$.

Problem 367. What postage do you think we can make with five and six cent stamps? Is there a number N such that if $n \geq N$, then we can make n cents worth of postage?

You probably see that we can make n cents worth of postage as long as n is at least 20. However you didn't try to make 26 cents in postage by working with 25 cents; rather you saw that you could get 20 cents and then add six cents to that to get 26 cents. Thus if we want to prove by induction that we are right that if $n \geq 20$, then we can make n cents worth of postage, we are going to have to use the strong version of the principle of mathematical induction.

We know that we can make 20 cents with four five-cent stamps. Now we let k be a number greater than 20, and assume that it is possible to make any amount between 20 and $k - 1$ cents in postage with five and six cent stamps. Now if k is less than 25, it is 21, 22, 23, or 24. We can make 21 with three fives and one six. We can make 22 with two fives and two sixes, 23 with one five and three sixes, and 24 with four sixes. Otherwise $k - 5$ is between 20 and $k - 1$ (inclusive) and so by our inductive hypothesis, we know that $k - 5$ cents can be made with five and six cent stamps, so with one more five cent stamp, so can k cents. Thus by the (strong) principle of mathematical induction, we can make n cents in stamps with five and six cent stamps for each $n \geq 20$.

Some people might say that we really had five base cases, $n = 20, 21, 22, 23$, and 24, in the proof above and once we had proved those five consecutive base cases, then we could reduce any other case to one of these base cases by successively subtracting 5. That is an appropriate way to look at the proof. A logician would say that it is also the case that, for example, by proving we could make 22 cents, we also proved that if we can make 20 cents and 21 cents in stamps, then we could also make 22 cents. We just didn't bother to use the assumption that we could make 20 cents and 21 cents! So long as one point of view or the other satisfies you, you are ready to use this kind of argument in proofs.

Problem 368. A number greater than one is called prime if it has no factors other than itself and one. Show that each positive number is either a prime or a power of a prime or a product of powers of prime numbers.

Problem 369. Show that the number of prime factors of a positive number $n \geq 2$ is less than or equal to $\log_2 n$. (If a prime occurs to the kth power in a factorization of n, you can consider that power as k prime factors.) (There is a way to do this by induction and a way to do it without induction. It would be ideal to find both ways.)

Problem 370. One of the most powerful statements in elementary number theory is Euclid's Division Theorem.[a] This states that if m and n are positive integers, then there are unique nonnegative intergers q and r with $0 \leq r < n$, such that $m = nq + r$. The number q is called the quotient and the number r is called the remainder. In computer science it is common to denote r by $m \mod n$. In elementary school you learned how to use long division to find q and r. However, it is unlikely that anyone ever proved for you that for any pair of positive intgers, m and n, there is such a pair of nonnegative numbers q and r. You now have the tools needed to prove this. Do so. (h)

---

[a]In a curious twist of language, mathematicians have long called The Division Algorithm or Euclid's Division Algorithm. However as computer science has grown in importance, the word algorithm has gotten a more precise definition: an algorithm is now a method to do something. There is a method (in fact there are more than one) to get the q and r that Euclid's Division Theorem gives us, and computer scientists would call these methods algorithms. Your author has chosen to break with mathematical tradition and restrict his use of the word algorithm to the more precise interpretation as a computer scientist probably would. We aren't giving a method here, so this is why the name used here is "Euclid's Division Theorem."

Appendix C

Exponential Generating Functions

C.1 Indicator Functions

When we introduced the idea of a generating function, we said that the formal sum

$$\sum_{i=0}^{n} a_i x^i$$

may be thought of as a convenient way to keep track of the sequence a_i. We then did quite a few examples that showed how combinatorial properties of arrangements counted by the coefficients in a generating function could be mirrored by algebraic properties of the generating functions themselves. The monomials x^i are called **indicator polynomials**. (They indicate the position of the coefficient a_i.) One example of a generating function is given by

$$(1+x)^n = \sum_{i=0}^{\infty} \binom{n}{i} x^i.$$

Thus we say that $(1+x)^n$ is the generating function for the binomial coefficients $\binom{n}{i}$. The notation tells us that we are assuming that only i varies in the sum on the right, but that the equation holds for each fixed integer n. This is implicit when we say that $(1+x)^n$ is the generating function for $\binom{n}{i}$, because we haven't written i anywhere in $(1+x)^n$, so it is free to vary.

Another example of a generating function is given by

$$x^{\underline{n}} = \sum_{i=0}^{\infty} s(n,i) x^i.$$

Thus we say that $x^{\underline{n}}$ is the generating function for the Stirling numbers of the first kind, $s(n,i)$. There is a similar equation for Stirling numbers of the second kind, namely

$$x^n = \sum_{i=0}^{\infty} S(n,i) x^{\underline{i}}.$$

151

However with our previous definition of generating functions, this equation would not give a generating function for the Stirling numbers of the second kind, because $S(n,i)$ is not the coefficient of x^i. If we were willing to consider the falling factorial powers $x^{\underline{i}}$ as indicator polynomials, then we could say that x^n is the generating function for the numbers $S(n,i)$ relative to these indicator polynomials. This suggests that perhaps different sorts of indicator polynomials go naturally with different sequences of numbers.

The binomial theorem gives us yet another example.

○ **Problem 371.** Write $(1+x)^n$ as a sum of multiples of $\frac{x^i}{i!}$ rather than as a sum of multiples of x^i.

This example suggests that we could say that $(1+x)^n$ is the generating function for the falling factorial powers $n^{\underline{i}}$ relative to the indicator polynomials $\frac{x^i}{i!}$. In general, a sequence of polynomials is called a family of **indicator polynomials** if there is one polynomial of each nonnegative integer degree in the sequence. Those familiar with linear algebra will recognize that this says that a family of indicator polynomials form a basis for the vector space of polynomials. This means that each polynomial way can be expressed as a sum of numerical multiples of indicator polynomials in one and only one way. One could use the language of linear algebra to define indicator polynomials in an even more general way, but a definition in such generality would not be useful to us at this point.

C.2 Exponential Generating Functions

We say that the expression $\sum_{i=0}^{\infty} a_i \frac{x^i}{i!}$ is the **exponential generating function** for the sequence a_i. It is standard to use **EGF** as a shorthand for exponential generating function. In this context we call the generating function $\sum_{i=0}^{n} a_i x^i$ that we originally studied the **ordinary generating function** for the sequence a_i. You can see why we use the term exponential generating function by thinking about the exponential generating function (EGF) for the all ones sequence,

$$\sum_{i=0}^{\infty} 1 \frac{x^i}{i!} = \sum_{i=0}^{\infty} \frac{x^i}{i!} = e^x,$$

which we also denote by $\exp(x)$. Recall from calculus that the usual definition of e^x or $\exp(x)$ involves limits at least implicitly. We work our way around that by defining e^x to be the power series $\sum_{i=0}^{\infty} \frac{x^i}{i!}$.

○ **Problem 372.** Find the EGF (exponential generating function) for the sequence $a_n = 2^n$. What does this say about the EGF for the number of subsets of an n-element set?

- **Problem 373.** Find the EGF (exponential generating function) for the number of ways to paint the n streetlight poles that run along the north side of Main Street in Anytown, USA using four colors.

Problem 374. For what sequence is $\frac{e^x - e^{-x}}{2} = \cosh x$ the EGF (exponential generating function)?

- **Problem 375.** For what sequence is $\ln(\frac{1}{1-x})$ the EGF? ($\ln(y)$ stands for the natural logarithm of y. People often write $\log(y)$ instead.) Hint: Think of the definition of the logarithm as an integral, and don't worry at this stage whether or not the usual laws of calculus apply, just use them as if they do! We will then define $\ln(1-x)$ to be the power series you get. [a]

  ---
  [a] It is possible to define the derivatives and integrals of power series by the formulas

  $$\frac{d}{dx} \sum_{i=0}^{\infty} b_i x^i = \sum_{i=1}^{\infty} i b_i x^{i-1}$$

  and

  $$\int_0^x \sum_{i=0}^{\infty} b_i x^i = \sum_{i=0}^{\infty} \frac{b_i}{i+1} x^{i+1}$$

  rather than by using the limit definitions from calculus. It is then possible to prove that the sum rule, product rule, etc. apply. (There is a little technicality involving the meaning of composition for power series that turns into a technicality involving the chain rule, but it needn't concern us at this time.)

- **Problem 376.** What is the EGF for the number of permutations of an n-element set?

⇒ - **Problem 377.** What is the EGF for the number of ways to arrange n people around a round table? Try to find a recognizable function represented by the EGF. Notice that we may think of this as the EGF for the number of permutations on n elements that are cycles. (h)

⇒ - **Problem 378.** What is the EGF $\sum_{n=0}^{\infty} p_{2n} \frac{x^{2n}}{(2n)!}$ for the number of ways p_{2n} to pair up $2n$ people to play a total of n tennis matches (as in Problems 12 and 44)? (h)

○ **Problem 379.** What is the EGF for the sequence $0, 1, 2, 3, \ldots$? You may think of this as the EFG for the number of ways to select one element from an n element set. What is the EGF for the number of ways to select two elements from an n-element set?

• **Problem 380.** What is the EGF for the sequence $1, 1, \ldots, 1, \ldots$? Notice that we may think of this as the EGF for the number of identity permutations on an n-element set, which is the same as the number of permutations of n elements that are products of 1-cycles, or as the EGF for the number of ways to select an n-element set (or, if you prefer, an empty set) from an n-element set. As you may have guessed, there are many other combinatorial interpretations we could give to this EGF.

○ **Problem 381.** What is the EGF for the number of ways to select n distinct elements from a one-element set? What is the EGF for the number of ways to select a positive number n of elements from a one element set? Hint: When you get the answer you will either say "of course," or "this is a silly problem." (h)

• **Problem 382.** What is the EGF for the number of partitions of a k-element set into exactly one block? (Hint: is there a partition of the empty set into exactly one block?)

• **Problem 383.** What is the EGF for the number of ways to arrange k books on one shelf (assuming they all fit)? What is the EGF for the number of ways to arrange k books on a fixed number n of shelves, assuming that all the books can fit on any one shelf? (Remember Problem 122.)

C.3 Applications to recurrences.

We saw that ordinary generating functions often play a role in solving recurrence relations. We found them most useful in the constant coefficient case. Exponential generating functions are useful in solving recurrence relations where the coefficients involve simple functions of n, because the $n!$ in the denominator can cancel out factors of n in the numerator.

○ **Problem 384.** Consider the recurrence $a_n = na_{n-1} + n(n-1)$. Multiply both sides by $\frac{x^n}{n!}$, and sum from $n = 2$ to ∞. (Why do we sum from $n = 2$ to infinity instead of $n = 1$ or $n = 0$?) Letting $y = \sum_{i=0}^{\infty} a_i x^i$, show that the left-hand side of the equation is $y - a_0 - a_1 x$. Express the right hand side in terms of y, x, and e^x. Solve the resulting equation for y and use the result to get an equation for a_n. (A finite summation is acceptable in your answer for a_n.)

⇒ · **Problem 385.** The telephone company in a city has n subscribers. Assume a telephone call involves exactly two subscribers (that is, there are no calls to outside the network and no conference calls), and that the configuration of the telephone network is determined by which pairs of subscribers are talking. Notice that we may think of a configuration of the telephone network as a permutation whose cycle decomposition consists entirely of one-cycles and two-cycles, that is, we may think of a configuration as an involution in the symmetric group S_n.

   (a) Give a recurrence for the number c_n of configurations of the network. (Hint: Person n is either on the phone or not.)

   (b) What are c_0 and c_1?

   (c) What are c_2 through c_6?

⇒ · **Problem 386.** Recall that a **derangement** of $[n]$ is a permutation of $[n]$ that has no fixed points, or equivalently is a way to pass out n hats to their n different owners so that nobody gets the correct hat. Use d_n to stand for the number of derangements of $[n]$. We can think of derangement of $[n]$ as a list of 1 through n so that i is not in the ith place for any n. Thus in a derangement, some number k different from n is in position n. Consider two cases: either n is in position k or it is not. Notice that in the second case, if we erase position n and replace n by k, we get a derangement of $[n-1]$. Based on these two cases, find a recurrence for d_n. What is d_1? What is d_2? What is d_0? What are d_3 through d_6?

C.3.1 Using calculus with exponential generating functions

⇒ · **Problem 387.** Your recurrence in Problem 385 should be a second order recurrence.

   (a) Assuming that the left hand side is c_n and the right hand side involves c_{n-1} and c_{n-2}, decide on an appropriate power of x divided by an appropriate factorial by which to multiply both sides of the recurrence. Using the fact that the derivative of $\frac{x^n}{n!}$ is $\frac{x^{n-1}}{(n-1)!}$, write down a differential equation for the EGF $T(x) = \sum_{i=0}^{\infty} c_i \frac{x^i}{i!}$. Note that it makes sense to substitute 0 for x in $T(x)$. What is $T(0)$? Solve your differential equation to find an equation for $T(x)$.

(b) Use your EGF to compute a formula for c_n. (h)

⇒ • **Problem 388.** Your recurrence in Problem 386 should be a second order recurrence.

(a) Assuming that the left-hand side is d_n and the right hand side involves d_{n-1} and d_{n-2}, decide on an appropriate power of x divided by an appropriate factorial by which to multiply both sides of the recurrence. Using the fact that the derivative of $\frac{x^n}{n!}$ is $\frac{x^{n-1}}{(n-1)!}$, write down a differential equation for the EGF $D(x) = \sum_{i=0}^{\infty} d_i \frac{x^i}{i!}$. What is $D(0)$? Solve your differential equation to find an equation for $D(x)$.

(b) Use the equation you found for $D(x)$ to find an equation for d_n. Compare this result with the one you computed by inclusion and exclusion.

C.4 The Product Principle for EGFs

One of our major tools for ordinary generating functions was the product principle. It is thus natural to ask if there is a product principle for exponential generating functions. In Problem 383 you likely found that the EGF for the number of ways of arranging n books on one shelf was exactly the same as the EGF for the number of permutations of $[n]$, namely $\frac{1}{1-x}$ or $(1-x)^{-1}$. Then using our formula from Problem 122 and the generating function for multisets, you probably found that the EGF for number of ways of arranging n books on some fixed number m of bookshelves was $(1-x)^{-m}$. Thus the EGF for m shelves is a product of m copies of the EGF for one shelf.

○ **Problem 389.** In Problem 373 what would the exponential generating function have been if we had asked for the number of ways to paint the poles with just one color of paint? With two colors of paint? What is the relationship between the EGF for painting the n poles with one color of paint and the EGF for painting the n poles with four colors of paint? What is the relationship between the EGF for painting the n poles with two colors of paint and the EGF for painting the poles with four colors of paint?

In Problem 385 you likely found that the EGF for the number of network configurations with n customers was $e^{x+x^2/2} = e^x \cdot e^{x^2/2}$. In Problem 380 you saw that the generating function for the number of permutations on n elements that are products of one cycles was e^x, and in Problem 378 you likely found that the EGF for the number of tennis pairings of $2n$ people, or equivalently, the number of permutations of $2n$ objects that are products of n two-cycles is $e^{x^2/2}$.

- **Problem 390.** What can you say about the relationship among the EGF for the number of permutations that are products of disjoint two-cycles and one-cycles, i.e., are involutions, the exponential generating function for the number of permutations that are the product of disjoint two-cycles only and the generating function for the number of permutations that are the product of disjoint one cycles only (these are identity permutations on their domain)?

In Problem 388 you likely found that the EGF for the number of permutations of $[n]$ that are derangements is $\frac{e^{-x}}{1-x}$. But every permutation is a product of derangements and one cycles, because the permutation that sends i to i is a one-cycle, so that when you factor a permutation as a product of disjoint cycles, the cycles of size greater than one multiply together to give a derangement, and the elements not moved by the permutation are one-cycles.

- **Problem 391.** If we multiply the EGF for derangements times the EGF for the number of permutations whose cycle decompositions consist of one-cycles only, what EGF do we get? for what set of objects have we found the EGF? (h)

We now have four examples in which the EGF for a sequence or a pair of objects is the product of the EGFs for the individual objects making up the sequence or pair.

- **Problem 392.** What is the coefficient of $\frac{x^n}{n!}$ in the product of two EGFs $\sum_{i=0}^{\infty} a_i \frac{x^i}{i!}$ and $\sum_{j=0}^{\infty} b_j \frac{x^j}{j!}$? (A summation sign is appropriate in your answer.) (h)

In the case of painting streetlight poles in Problem 389, let us examine the relationship between the EGF for painting poles with two colors, the EGF for painting the poles with three colors, and the EGF for painting poles with five colors, e^{5x}. To be specific, the EGF for painting poles red and white is e^{2x} and the EGF for painting poles blue, green, and yellow is e^{3x}. To decide how to paint poles with red, white, blue, green, and yellow, we can decide which set of poles is to be painted with red and white, and which set of poles is to be painted with blue, green, and yellow. Notice that the number of ways to paint a set of poles with red and white depends only on the size of that set, and the number of ways to paint a set of poles with blue, green, and yellow depends only on the size of that set.

- **Problem 393.** Suppose that a_i is the number of ways to paint a set of i poles with red and white, and b_j is the number of ways to paint a set of j poles with blue, green, and yellow. In how many ways may we take a set N of n poles, divide it up into two sets I and J (using i to stand for the size of I and j to stand for the size of the set J, and allowing i and j to vary) and paint the poles in I red and white and the poles in J blue, green, and yellow? (Give your answer in terms of a_i and b_j. Don't figure out formulas for a_i and b_j to use in your answer; that will make it harder to get the point of the problem!) How does this relate to Problem 392?

Problem 393 shows that the formula you got for the coefficient of $\frac{x^n}{n!}$ in the product of two EGFs is the formula we get by splitting a set N of poles into two parts and painting the poles in the first part with red and white and the poles in the second part with blue, green, and yellow. More generally, you could interpret your result in Problem 392 to say that the coefficient of $\frac{x^n}{n!}$ in the product $\sum_{i=0}^{\infty} a_i \frac{x^i}{i!} \sum_{j=0}^{\infty} b_j \frac{x^j}{j!}$ of two EGFs is the sum, over all ways of splitting a set N of size n into an ordered pair of disjoint sets I and J, of the product $a_{|I|} b_{|J|}$.

There seem to be two essential features that relate to the product of exponential generating functions. First, we are considering **structures** that consist of a set and some additional mathematical construction on or relationship among the elements of that set. For example, our set might be a set of light poles and the additional construction might be a coloring function defined on that set. Other examples of additional mathematical constructions or relationships on a set could include a permutation of that set; in particular an involution or a derangement, a partition of that set, a graph on that set, a connected graph on that set, an arrangement of the elements of that set around a circle, or an arrangement of the elements of that set on the shelves of a bookcase. In fact a set with no additional construction or arrangement on it is also an example of a structure. Its additional construction is the empty set! When a structure consists of the set S plus the additional construction, we say the structure *uses* S. What all the examples we have mentioned in our earlier discussion of exponential generating functions have in common is that the number of structures that use a given set is determined by the size of that set. We will call a family $\mathcal{F}$ of structures a **species** of structures on subsets of a set X if structures are defined on finite subsets of X and if the number of structures in the family using a finite set S is finite and is determined by the size of S (that is, if there is a bijection between subsets S and T of X, the number of structures in the family that use S equals the number of structures in the family that use T). We say a structure is an $\mathcal{F}$-**structure** if it is a member of the family $\mathcal{F}$.

- **Problem 394.** In Problem 383, why is the family of arrangements of set of books on a single shelf (assuming they all fit) a species?

- **Problem 395.** In Problem 385, why is the family of people actually making phone calls (assuming nobody is calling outside the telephone network) at any given time, with the added relationship of who is calling whome, a species? Why is the family of sets of people who are not using their phones a species (with no additional construction needed)?

The second essential feature of our examples of products of EGFs is that products of EGFs seem to count structures on ordered pairs of two disjoint sets (or more generally on k-tuples of mutually disjoint sets). For example, we can determine a five coloring of a set S by partitioning it in all possible ways into two sets and coloring the first set in the pair with our first two colors and our second pair with the last three colors. Or we can partition our set in all possible ways into five parts and color part i with our ith color. We don't have to do the same thing to each part of our partition; for example, we could define a derangement on one part and an identity permutation on the other; this defines a permutation on the set we are partitioning, and we have already noted that every permutation arises in this way.

Our combinatorial interpretation of EGFs will involve assuming that the coefficient of $\frac{x^i}{i!}$ counts the number of structures on a particular set of of size i in a species of structures on subsets of a set X. Thus in order to give an interpretation of the product of two EGFs we need to be able to think of ordered pairs of structures on disjoint sets or k-tuples of structures on disjoint sets as structures themselves. Thus given a structure on a set S and another structure on a disjoint set T, we define the ordered pair of structures (which is a mathematical construction!) to be a structure on the set $S \cup T$. We call this a **pair structure** on $S \cup T$. We can get many structures on a set $S \cup T$ in this way, because $S \cup T$ can be divided into many other pairs of disjoint sets. In particular, the set of pair structures whose first structure comes from $\mathcal{F}$ and whose second element comes from $\mathcal{G}$ is denoted by $\mathcal{F} \cdot \mathcal{G}$.

- **Problem 396.** Show that if $\mathcal{F}$ and $\mathcal{G}$ are species of structures on subsets of a set X, then the pair of structures of $\mathcal{F} \cdot \mathcal{G}$ for a species of structures

Given a species $\mathcal{F}$ of structures, the number of structures using any particular set of size i is the same as the number of structures in the family using any other set of size i. We can thus define the exponential generating function (EGF) for the family as the power series $\sum_{i=1}^{\infty} a_i \frac{x^i}{i!}$, where a_i is the number of structures of $\mathcal{F}$ that use one particular set of size i. In Problems 372, 373, 376, 377, 378, 380, 382, 383, 387, and 388, we were computing EGFs for species of subsets of some set.

- **Problem 397.** If $\mathcal{F}$ and $\mathcal{G}$ are species of subsets of X, how is the EGF for $\mathcal{F} \cdot \mathcal{G}$ related to the EGFs for F and G? Prove you are right. (h)

Problem 398. Without giving the proof, how can you compute the EGF $f(x)$ for the number of structures using a set of size n in the species $\mathcal{F}_1 \cdot \mathcal{F}_2 \cdots \mathcal{F}_k$ of structures on k-tuples of subsets of X from the EGFs $f_i(x)$ for $\mathcal{F}_i$ for each i from 1 to k? (Here we are using the natural extension of the idea of the pair of structures to the idea of a k-tuple structure.)

Theorem C.4.1. *If $\mathcal{F}_1, \mathcal{F}_2, \ldots, \mathcal{F}_n$ are species of subsets of the set X and $\mathcal{F}_i$ has EGF $f_i(x)$, then the family of k-tuple structures $\mathcal{F}_1 \cdot \mathcal{F}_2 \cdots \mathcal{F}_n$ has EGF $\prod_{i=1}^{n} f_i(x)$.*

We call Theorem C.4.1 the **product principle for exponential generating functions**. We give two corollaries; the proof of the second is not immediate though not particular difficult.

Corollary C.4.2. *If $\mathcal{F}$ is a species of structures on subsets of X and $f(x0)$ is the EGF for $\mathcal{F}$, then $f(x)^k$ is the EGF for the k-tuple structure on k-tuples of $\mathcal{F}$-structures using disjoint subsets of X.*

Our next corollary uses the idea of a k-set structure. Suppose we have a species $\mathcal{F}$ of structures on nonempty subsets of X, that is, a species of structures which assigns no structures to the empty set. Then we can define a new species $\mathcal{F}^{(k)}$ of structures, called "k-set structures," using nonempty subsets of X. Given a fixed positive integer k, a k-**set structure** on a subset Y of X consists of a k-element set of nonempty disjoint subsets of X whose union is Y and an assignment of an $\mathcal{F}$-structure to each of the disjoint subsets. This is a species on the set of subsets of X; the subset used by a k-set structure is the union of the sets of the structure. To recapitulate, the set of k-set structures on a subset Y of X is the set of all possible assignments of $\mathcal{F}$-structures to k nonempty disjoint sets whose union is Y. (You can also think of the k-set structures as a family of structures defined on blocks of partitions of subsets of X into k blocks.)

Corollary C.4.3. *If $\mathcal{F}$ is a species of structures on nonempty subsets of X and $f(x)$ is the EGF for $\mathcal{F}$, then for each positive integer k, $\frac{f(x)^k}{k!}$ is the EGF for the family $\mathcal{F}^{(k)}$ of k-set structures on subsets of X*

Problem 399. Prove Corollary C.4.3. (h)

- **Problem 400.** Use the product principle for EGFs to explain the results of Problems 390 and Problem 391.

- **Problem 401.** Use the general product principle for EGFs or one of its corollaries to explain the relationship between the EGF for painting streetlight poles in only one color and the EGF for painting streetlight poles in 4 colors in Problems 373 and Problem 389. What is the EGF for the number p_n of ways to paint n streetlight poles with some fixed number k of colors of paint.

- **Problem 402.** Use the general product principle for EGFs or one of its corollaries to explain the relationship between the EGF for arranging books on one shelf and the EGF for arranging books on n shelves in Problem 383.

⇒ **Problem 403.** (Optional) Our very first example of exponential generating functions used the binomial theorem to show that the EGF for k-element permutations of an n element set is $(1+x)^n$. Use the EGF for k-element permutations of a one-element set and the product principle to prove the same thing. Hint: Review the alternate definition of a function in Section 3.1.2. (h)

Problem 404. What is the EGF for the number of ways to paint n streetlight poles red, white blue, green and yellow, assuming an even number of poles must be painted green and an even number of poles must be painted yellow? Give a formula for the number of ways to paint n poles. (Don't forget the factorial!) (h)

⇒ · **Problem 405.** What is the EGF for the number of functions from an n-element set onto a one-element set? (Can there be any functions from the empty set onto a one-element set?) What is the EGF for the number c_n of functions from an n-element set onto a k element set (where k is fixed)? Use this EGF to find an explicit expression for the number of functions from a k-element set onto an n-element set and compare the result with what you got by inclusion and exclusion.

⇒ · **Problem 406.** In Problem 142 You showed that the Bell Numbers B_n satisfy the equation $B_{n+1} = \sum_{k=0}^{n} \binom{n}{k} B_{n-k}$ (or a similar equation for B_n.) Multiply both sides of this equation by $\frac{x^n}{n!}$ and sum from $n = 0$ to infinity. On the left hand side you have a derivative of a certain EGF we might call $B(x)$. On the right hand side, you have a product of two EGFs, one of which is $B(x)$. What is the other one? What differential equation involving $B(x)$ does this give you. Solve the differential equation for $B(x)$. This is the EGF for the Bell numbers!.

⇒ **Problem 407.** Prove that $n2^{n-1} = \sum_{k=1}^{n} \binom{n}{k} k$ by using EGFs. (h)

• **Problem 408.** In light of Problem 382, why is the EGF for the Stirling numbers $S(n,k)$ of the second kind not $(e^x - 1)^n$? What is it equal to instead?

C.5 The Exponential Formula

Exponential generating functions turn out to be quite useful in advanced work in combinatorics. One reason why is that it is often possible to give a combinatorial interpretation to the composition of two exponential generating functions. In particular, if $f(x) = \sum_{i=0}^{n} a_i \frac{x^i}{i!}$ and $g(x) = \sum_{j=1}^{\infty} b_j \frac{x^j}{j!}$, it makes sense to form the composition $f(g(x))$ because in so doing we need add together only finitely many terms in order to find the coefficient of $\frac{x^n}{n!}$ in $f(g(x))$ since in the EGF $g(x)$ the dummy variable j starts at 1. Since our study of combinatorial structures has not been advanced enough to give us applications of a general formula for the composition of the EGF, we will not give here the combinatorial interpretation of this composition. However we have seen some examples where one particular composition can be applied. Namely, if $f(x) = e^x = \exp(x)$, then $f(g(x)) = \exp(g(x))$ is well defined when $b_0 = 0$. We have seen three examples in which an EGF is $e^{f(x)}$ where $f(x)$ is another EGF. There is a fourth example in which the exponential function is slightly hidden.

• **Problem 409.** If $f(x)$ is the EGF for the number of partitions of an n-set into one block, and $g(x)$ is the EGF for the total number of partitions of an n-element set, that is, for the Bell numbers B_n, how are the two generating functions related?

• **Problem 410.** Let $f(x)$ be the EGF for the number of permutations of an n-element set with one cycle of size one or two. What is $f(x)$? What is the EGF $g(x)$ for the number of permutations of an n-element set all of whose cycles have size one or two, that is, the number of involutions in S_n? How are these two exponential generating functions related?

⇒ • **Problem 411.** Let $f(x)$ be the EGF for the number of permutations of an n-element set that have exactly one two-cycle and no other cycles. Let $g(x)$ be the EGF for the number of permutations which are products of two-cycles only, that is, for tennis pairings. (That is, they are not a product of two-cycles and a nonzero number of one-cycles). What is $f(x)$? What is $g(x)$? How are these to exponential generating functions related?

- **Problem 412.** Let $f(x)$ be the EGF for the number of permutations of an n-element set that have exactly one cycle. (This is the same as the EGF for the number of ways to arrange n people around a round table.) Let $g(x)$ be the EGF for the total number of permutations of an n-element set. What is $f(x)$? What is $g(x)$? How are $f(x)$ and $g(x)$ related?

There was one element that our last four problems had in common. In each case our EGF $f(x)$ involved the number of structures of a certain type (partitions, telephone networks, tennis pairings, permutations) that used only one set of an appropriate kind. (That is, we had a partition with one part, a telephone network consisting either of one person or two people connected to each other, a tennis pairing of one set of two people, or a permutation with one cycle.) Our EGF $g(x)$ was the number of structures of the same "type" (we put type in quotation marks here because we don't plan to define it formally) that could consist of any number of sets of the appropriate kind. Notice that the order of these sets was irrelevant. For example we don't order the blocks of a partition and a product of disjoint cycles is the same no matter what order we use to write down the product. Thus we were relating the EGF for structures which were somehow "building blocks" to the EGF for structures which were sets of building blocks. For a reason that you will see later, it is common to call the building blocks **connected** structures. Notice that our connected structures were all based on nonempty sets, so we had no connected structures whose value was the empty set. Thus in each case, if $f(x) = \sum_{i=0}^{\infty} a_i \frac{x^i}{i!}$, we would have $a_0 = 0$. The relationship between the EGFs was always $g(x) = e^{f(x)}$. We now give a combinatorial explanation for this relationship.

- **Problem 413.** Suppose that $\mathcal{F}$ is a species of structures of a set X with no structures on the empty set. Let $f(x)$ be the EGF for $\mathcal{F}$.

    (a) In the power series
    $$e^{f(x)} = 1 + f(x) + \frac{f(x)^2}{2!} + \cdots + \frac{f(x)^k}{k!} + \cdots = \sum_{k=0}^{\infty} \frac{f(x)^k}{k!},$$
    what does Corollary C.4.3 tell us about the coefficient of $\frac{x^n}{n!}$ in $\frac{f(x)^k}{k!}$?

    (b) What does the coefficient of $\frac{x^n}{n!}$ in $e^{f(x)}$ count?

In Problem 413 we proved the following theorem, which is called the **exponential formula**.

Theorem C.5.1. *Suppose that $\mathcal{F}$ is a species of structures on subsets of a set X with no structures on the empty set. Let $f(x)$ be the EGF for $\mathcal{F}$. Then the coefficient of $\frac{x^n}{n!}$ in $e^{f(x)}$ is the number of sets of structures on disjoint sets whose union is a particular set of size n.*

Let us see how the exponential formula applies to the examples in Problems 409, 410, 411 and 412. In Problem 382 our family $\mathcal{F}$ should consist of one-block partitions of finite subsets of a set, say the set of natural numbers. Since a partition of a set is a set of blocks whose union is S, a one-block partition whose block is B is the set $\{B\}$. Then any nonempty finite subset of of the positive integers is the value of exactly one structure in $\mathcal{F}$. (There is no one-block partition of the empty set, so we have no structures using the empty set.) As you showed in Problem 382 the generating function for partitions with just one block is $e^x - 1$. Thus by the exponential formula, $\exp(e^x - 1)$ is the EGF for sets of subsets of the positive integers whose values are disjoint sets whose union is any particular set N of size n. This set of disjoint sets partitions the set N. Thus $\exp(e^x - 1)$ is the EGF for partitions of sets of size n. (As we wrote our description, it is the EGF for partitions of n-element subsets of the positive integers, but any two n-element sets have the same number of partitions.) In other words, $\exp(e^x - 1)$ is the exponential generating function for the Bell numbers B_n.

- **Problem 414.** Explain how the exponential formula proves the relationship we saw in Problem 412.

- **Problem 415.** Explain how the exponential formula proves the relationship we saw in Problem 411.

- **Problem 416.** Explain how the exponential formula proves the relationship we saw in Problem 410.

- **Problem 417.** In Problem 373 we saw that the generating function for the number of ways to use five colors of paint to paint n light poles along the north side of Main Street in Anytown was e^{4x}. We should expect an explanation of this EGF using the exponential formula. Let $\mathcal{F}$ be the family of all one-element sets of light poles with the additional construction of an ordered pair consisting of a light pole and a color. Thus a given light pole occurs in five ordered pairs. Put no structures on any other finite set. Show that this is a species of structures on the finite subsets of the positive integers. What is the exponential generating function $f(x)$ for $\mathcal{F}$? Assuming that there is no upper limit on the number of light poles, what subsets of S does the exponential formula tell us are counted by the coefficient of x^n in $e^{f(x)}$? How do the sets being counted relate to ways to paint light poles?

One of the most spectacular applications of the exponential formula is also the reason why, when we regard a combinatorial structure as a set of building block structures, we call the building block structures **connected**. In Chapter 2 we introduced the idea of a connected graph and in Problem 104 we saw examples of graphs which were connected and were not connected. A subset C of the vertex set of a graph is called a **connected component** of the graph if

- every vertex in C is connected to every other vertex in that set by a walk whose vertices lie in C, and

- no other vertex in the graph is connected by a walk to any vertex in C.

In Problem 241 we showed that each connected component of a graph consists of a vertex and all vertices connected to it by walks in the graph.

- **Problem 418.** Show that every vertex of a graph lies in one and only one connected component of a graph. (Notice that this shows that the connected components of a graph form a partition of the vertex set of the graph.)

- **Problem 419.** Explain why no edge of the graph connects two vertices in different connected components.

- **Problem 420.** Explain why it is that if C is a connected component of a graph and E' is the set of all edges of the graph that connect vertices in C, then the graph with vertex set C and edge set E' is a connected graph. We call this graph a **connected component graph** of the original graph.

The last sequence of problems shows that we may think of any graph as the set of its connected component graphs. (Once we know them, we know all the vertices and all the edges of the graph). Notice that a graph is connected if and only if it has exactly one connected component. Since the connected components form a partition of the vertex set of a graph, the exponential formula will relate the EGF for the number of connected graphs on n vertices with the EGF for the number of graphs (connected or not) on n vertices. However because we can draw as many edges as we want between two vertices of a graph, there are infinitely many graphs on n vertices, and so the problem of counting them is uninteresting. We can make it interesting by considering **simple graphs**, namely graphs in which each edge has two distinct endpoints and no two edges connect the same two vertices. It is because connected graphs form the building blocks for viewing all graphs as sets of connected components that we refer to the building blocks for structures counted by the EGF in the exponential formula as connected structures.

⇒ · **Problem 421.** Suppose that $f(x) = \sum_{n=0}^{\infty} c_n \frac{x^n}{n!}$ is the exponential generating function for the number of simple connected graphs on n vertices and $g(x) = \sum_{i=0}^{\infty} a_i \frac{x^i}{i!}$ is the exponential generating function for the number of simple graphs on i vertices. From this point onward, any use of the word graph means simple graph.

(a) Is $f(x) = e^{g(x)}$, is $f(x) = e^{g(x)-1}$, is $g(x) = e^{f(x)-1}$ or is $g(x) = e^{f(x)}$? (h)

(b) One of a_i and c_n can be computed by recognizing that a simple graph on a vertex set V is completely determined by its edge set and its edge set is a subset of the set of two element subsets of V. Figure out which it is and compute it. (h)

(c) Write $g(x)$ in terms of the natural logarithm of $f(x)$ or $f(x)$ in terms of the natural logarithm of $g(x)$.

(d) Write $\log(1+y)$ as a power series in y. (h)

(e) Why is the coefficient of $\frac{x^0}{0!}$ in $g(x)$ equal to one? Write $f(x)$ as a power series in $g(x) - 1$.

(f) You can now use the previous parts of the problem to find a formula for c_n that involves summing over all partitions of the integer n. (It isn't the simplest formula in the world, and it isn't the easiest formula in the world to figure out, but it is nonetheless a formula with which one could actually make computations!) Find such a formula. (h)

The point to the last problem is that we can use the exponential formula in reverse to say that if $g(x)$ is the generating function for the number of (nonempty) connected structures of size n in a given family of combinatorial structures and $f(x)$ is the generating function for all the structures of size n in that family, then not only is $f(x) = e^{g(x)}$, but $g(x) = \ln(f(x))$ as well. Further, if we happen to have a formula for either the coefficients of $f(x)$ or the coefficients of $g(x)$, we can get a formula for the coefficients of the other one!

C.6 Supplementary Problems

1. Use product principle for EGFs and the idea of coloring a set in two colors to prove the formula $e^x \cdot e^x = e^{2x}$.

2. Find the EGF for the number of ordered functions from a k-element set to an n-element set.

3. Find the EGF for the number of ways to string n distinct beads onto a necklace.

4. Find the exponential generating function for the number of broken permutations of a k-element set into n parts.

5. Find the EGF for the total number of broken permutations of a k-element set.

6. Find the EGF for the number of graphs on n vertices in which every vertex has degree 2.

7. Recall that a cycle of a permutation cannot be empty.
  (a) What is the generating function for the number of cycles on an even number of elements (i.e. permutations of an even number n of elements that form an n-cycle)? Your answer should not have a summation sign in it. Hint: If $y = \sum_{i=0}^{\infty} \frac{x^{2i}}{2i}$, what is the derivative of y?
  (b) What is the generating function for the number of permutations on n elements that are a product of even cycles?
  (c) What is the generating function for the number of cycles on an odd number of elements?
  (d) What is the generating function for the number of permutations on n elements that are a product of odd cycles?
  (e) How do the generating functions in parts b and d of this problem related to the generating function for all permutations on n elements?

Appendix D

Hints to Selected Problems

1. Answer the questions in Problem 2 for the case of five schools.

3. For each kind of bread, how many sandwiches are possible?

6. Try to solve the problem first with a two-scoop cone. (Look for an earlier problem that is analogous.) Then, for each two scoop cone, in how many ways can you put on a top scoop?

7.a. Ask yourself "how many choices do we have for $f(1)$?" Then ask how many choices we have for $f(2)$.

7.b. It may not be practical to write down rules for all the functions for this problem. But you could ask yourself how many choices we have for $f(1)$, how many we have for $f(2)$ and how many we have for $f(3)$.

7.c. If you are choosing a function f, how many choices do you have for $f(a)$? Then how many choices do you have for $f(b)$?

8.a. You know how to figure out in how many ways they could make a list of three flavors out of the twelve. But each set of three flavors can be listed in a number of different ways. Try to figure out in how many ways a set of three flavors can be listed, and then try to see how this helps you.

8.b. Try to break the problem up into cases you can solve by previous methods; then figure out how to get the answer to the problem by using these answers for the cases.

12.a. Suppose you have a list in alphabetical order of names of the members of the club. In how many ways can you pair up the first person on the list? In how many ways can you pair up the next person who isn't already paired up?

15. In how many ways may you assign the men to their rows? The women? Once a woman and a man have a row to share, in how many ways may they choose their seats?

18. Try applying the product principle in the case $n = 2$ and $n = 3$. How might you apply it in general?

19. Ask yourself if either the sum principle or product principle applies.

    Additional Hint: Remember that zero is a number.

20. Do you see an analogy between this problem and any of the previous problems?

26.a. For each part of this problem, think about how many arrows are allowed to enter a vertex representing a member of Y.

28. The problem is asking you to describe a one-to-one function from the set of binary representations of numbers between 0 and $2^n - 1$ onto the set of subsets of the set $[n]$. Write down these two sets for $n = 2$. They should both have four elements. The set of binary representations should contain the string 00. You could think of this as the instruction "take no ones and take no twos." In that context, what could you think of the string 11 as standing for? This should help you describe a function. Of course now you have to figure out how to show it is one-to-one and onto.

31. Starting with the row 1 8 28 56 70 56 28 8 1, put dots below it where the elements of row 9 should be. Then put dots below that where the elements of row 10 should be. Do the same for rows 11 and 12. Mark the dot where row 12 should appear. Now mark the dots you need in row 11 to compute the entry in column 3 of row 12. Now mark the dots you need in row 10 to compute the marked entries in row 11. Do the same for rows 9 and 8. Now you should be able to see what you need to do.

32.a. Begin by trying to figure out what the entries just above the diagonal of the rectangle are. After that, what other entries can you figure out?

32.b. See if you can figure out what the entries in column -1 have to be.

32.c. What does the sum of two consecutive values in row -1 have to be? Could this sum depend on which two consecutive values we take? Is there some value of row -1 that we could choose arbitrarily? Now what about row -2? Can we make arbitrary choices there? If so, how many can we make, and is their position arbitrary?

36. The first thing you need to decide is "What are the two sets whose elements we are counting?" Then it will be easier to think of a bijection between these two sets. It turns out that these two sets are sets of sets!

37. Ask yourself "What is a problem like this doing in the middle of a bunch of problems about counting subsets of a set? Is it related, or is it supposed to gives us a break from sets?"

38. The problem suggests that you think about how to get a list from a seating arrangement. Could every list of n distinct people come from a seating chart? How many lists of n distinct people are there? How many lists could we get from a given seating chart by taking different starting places?

    Additional Hint: For a different way of doing the problem, suppose that you have chosen one person, say the first one in a list of the people in alphabetical order by name. Now seat that person. Does it matter where they sit? In ways can you seat the remaining people? Does it matter where the second person in alphabetical order sits?

39.a. A block consists of all permutations of some subset $\{a_1, a_2, \ldots, a_k\}$ of S. How many permutations are there of the set $\{a_1, a_2, \ldots, a_k\}$?

39.c. What sets are listed, and how many times is each one listed if you take one list from each row of Table 1.2.8? How does this choice of lists give you the bijection in this special case?

39.d. You can make good use of the product principle here.

40.b. The coach is making a sequence of decisions. Can you figure out how many choices the coach has for each decision in the sequence?

40.c. As with any counting problem whose context does not suggest an approach, it is useful to ask yourself if you could decompose the problem into simpler parts by using either the sum or product principle.

43. How could we get a list of beads from a necklace?

    Additional Hint: When we cut the necklace and string it out on a table, there are 2n lists of beads we could get. Why is it $2n$ rather than n?

44.a. You might first choose the pairs of people. You might also choose to make a list of all the people and then take them by twos from the list.

44.b. You might first choose ordered pairs of people, and have the first person in each pair serve first. You might also choose to make a list of all the people and then take them by twos from the list in order.

45. It might be helpful to just draw some pictures of the possible configurations. There aren't that many.

47. Note that we must walk at least ten blocks, so ten is the smallest number of blocks possible. In how many of those ten blocks must we walk north?

48.b. In Problem 47 you saw that we had to make ten choices of north or east, choosing north four times.

48.c. This problem is actually a bit tricky. What happens to the answer if $i > m$ or $j > n$? Remember that paths go up or to the right.

49.a. Where can you go from $(0,0)$ in one step? In two steps? In any of these cases, what can you say about the sum of the coordinates of a point you can get to? Can you find any other relationship between the x- and y-coordinates of a point you can get to? For example, can you get to the point $(1,3)$?

49.c. How many choices do you have to make in order to choose a path?

50.c. In each part, each such sequence corresponds to a path that can't cross over (but may touch) a certain line.

51.b. Given a path from $(0,0)$ to (n,n) which touches or crosses the line $y = x+1$, how can you modify the part of the path from $(0,0)$ to the first touch of $y = x+1$ so that the modified path starts instead at $(-1,1)$? The trick is to do this in a systematic way that will give you your bijection.

51.c. A path either touches the line $y = x + 1$ or it doesn't. This partitions the set of paths into two blocks.

52.b. Look back at the definition of a Dyck Path and a Catalan Path.

52.c. What makes this part difficult is understanding how we are partitioning the paths. As an example, B_0 is the set of all paths that have no upsteps following the last absolute minimum. Can such a path have downsteps after the last absolute minimum? (The description we gave of B_0 is not succinct enough to be the answer to the second question of this part.) As another example B_1 is the set of all paths that have exactly one upstep and perhaps some downsteps after the last absolute minimum. Is it possible, though, for a path in B_1 to have any downsteps after the last absolute minimum? A path in B_2 has exactly two upsteps after its last absolute minimum. If is possible to have one downstep after the last absolute minimum, but it has to be in a special place. What place is that? Now to figure out how many parts our partition has, we need to know the maximum number of upsteps a path can have following its last absolute minimum. What is this maximum? It might help to draw some pictures with $n = 5$ or 6. In particular, is it possible that all upsteps occur after the last absolute minimum?

52.e. Using d for down and u for up, we could have $uudduuddudud$ as our Catalan path. Suppose that $i = 5$. The fifth upstep is the u in position 9. Thus $F = uudduudd$, $U = u$, and $B = dud$. Now BUF is $duduuudduudd$. This is a Dyck path that begins by going below the x-axis. The d's in positions 1 and 3 take the path to the y-coordinate -1. Then the y coordinate climbs to 2, goes back to 0, up to 2 again, and finally down to 0. So the absolute minimum is -1, and it occurs in the first and third position. There are five u's after the third positon. So this Dyck path is in the block B_5 of our partition. Now comes the crucial question. Why were there five u's after that last absolute minimum in position 3? Try with the same path and $i = 3$. Figure out why there are three u's after the last absolute minimum in the resulting path. All this discussion should explain why when $i = 5$, the set of all Catalan paths is mapped into the set B_5. Keeping $i = 5$ for a while, try to see why this correspondence between Catalan paths and B_5 is a bijection. Then, if you need to, do the same thing with $i = 3$. This should give you enough insight to do the general case.

55. What would the lower limit of the sum have to be for this problem to be a routine application of the binomial theorem?

56. What does the binomial theorem give you for $(x - y)^n$?

57. Consider $(x + y)^m (x + y)^n$.

    Additional Hint: What does $\binom{m+n}{k}$ count? What does $\binom{m}{i}\binom{m}{k-1}$ count?

58. For example when $n = 3$, we have $\binom{3}{0} = \binom{3}{3}$ and $\binom{3}{1} = \binom{3}{2}$. The number of subsets of even size is $\binom{3}{0} + \binom{3}{2}$ and the number of subsets of odd size is $\binom{3}{1} + \binom{3}{3}$, and the two sums can be paired off into equal terms. When we subtract the number of subsets of odd size from the number of subsets of even size, the pairing also gives us $\binom{3}{0} - \binom{3}{1} + \binom{3}{2} - \binom{3}{3} = 0$.

59. Take the derivative of something interesting.

61. To prove that each function from a set S of size n to a set of size less than n is not one-to-one, we must prove that regardless of the function f that we choose, there are always two elements, say x and y, such that $f(x) = f(y)$.

62. The previous exercise could help you prove that if f is one-to-one, then it is onto.

    Additional Hint: The sum principle can help you show that if f is an onto function, then f is one-to-one.

63. The statement of the generalized pigeonhole principle involves the number of elements in a block, so a counting principle is likely to help you.

64. You may choose a specific number for n if you want to. Notice that the last two digits of the powers of a prime other than two cannot represent an even number.

65. While this sounds like a pigeonhole principle problem, the ordinary pigeonhole principle doesn't guarantee three of something.

67. What usually makes it hard for students to start this problem is the fact that we just defined what $R(4,4)$ is, and not what it means for a number *not* to be $R(4,4)$. So to get started, try to write down what it means to say $R(4,4)$ is not 8. You will see that there are two things that can keep $R(4,4)$ from being 8. You need to figure out which one happens and explain why. One such explanation could involve the graph K_8.

68. Review Problem 65 and your solution of it.

69. Let a_i be the number of acquaintances of person i. Can you explain why the sum of the numbers a_i is even?

70. Often when there is a counter-example, there is one with a good deal of symmetry. (Caution: there is a difference between often and always!) One way to help yourself get a symmetric example, if there is one, is to put 8 vertices into a circle. Then, perhaps, you might draw green edges in some sort of regular fashion until it is impossible to draw another green edge between any two of the vertices without creating a green triangle.

71. In Problem 68 you showed that $R(4,3) \leq 10$. In Problem 70 you showed that $R(4,3) > 8$. Thus $R(4,3)$ is either 9 or 10. Deciding which is the case is just plain hard. But there is a relevant problem we have done that we haven't used yet.

72. We wish to prove that $\binom{n}{i} = \frac{n!}{i!(n-i)!}$. Mathematical induction allows us to assume that $\binom{n-1}{j} = \frac{(n-1)!}{j!(n-1-j)!}$ for every j between 0 and $n-1$. How does this put us into a position to use the Pascal relation? What special cases will be left over?

73. What sort of relationship do you know between values of the form $\binom{n}{i}$ and values of the form $\binom{n-1}{j}$?

75. We did something rather similar in our example of the inductive proof that a set with n elements has 2^n subsets. The work you did in a previous problem may be similar to part of what you need to do here.

76.a. This may look difficult because one can't decide in advance on whether to try to induct on m, on n, or on their sum. In some sense, it doesn't matter which you choose to induct on, though inducting on the sum would look more complicated. For most people inducting on n fits their way of working with exponents best.

76.b. Here it matters whether you choose to induct on m or n. However, it matters only in the sense that you need to use more tools in one case. In one case, you are likely to need the rule $(cd)^n = c^n d^n$ (, which we haven't proved. (However, you might be able to prove that by induction!) In either case, you may find part (a) handy.

79. We didn't explicitly say to use induction here, but, especially in this context, induction is a natural tool to try here. But we don't have a variable n to induct on. That means you have to choose one. So what do you think is most useful. The number of blocks in the partition? The size of the first block of the partition? The size of the set we are partitioning? Or something else?

80. Think about how you might have gone from the number of double decker cones to the number of triple decker cones in Problem 6.

81. Perhaps the first thing one needs to ask is why proving that if there are $\binom{m+n-2}{m-1}$ people in a room, then there are either at least m mutual acquaintances or at least n mutual strangers proves that $R(m,n)$ exists. Can you see why this tells us that there is some number R of people such that if R people are in a room, then there are m mutual acquaintances or n mutual strangers? And why does that mean the Ramsey Number exists?

    Additional Hint: Naturally it should come as no surprise that you will use double induction, and you can use either form. As you think about how to use induction, the Pascal relation will come to mind. This suggests that you want to make assumptions involving $\binom{m+n-3}{m-1}$ people in a room, or $\binom{m+n-3}{m-2}$ people in a room. Now you have to figure out what these assumptions are and how they help you prove the result! Recall that we have made progress before by choosing one person and asking whether this person is acquainted with at least some number of people or unacquainted with at least some other number of people.

82. One expects to need double induction again here. But only because of the location of the problem and because the sum looks like double induction. And those reasons aren't enough to mean you have to use double induction. If you had this result in hand already, then you could us it with double induction to give a second proof that Ramsey Numbers exist.

    Additional Hint: What you do need to show is that if there are $R(m-1,n) + R(m,n-1)$ people in a room, then there are either m mutual acquaintances or n mutual strangers. As with earlier problems, it helps to start with a person and think about the number of people with whom this person is acquainted or nonacquainted. The generalized pigeonhole principle tells you something about these numbers.

83.b. If you could find four mutual acquaintances, you could assume person 1 is among them. And by the generalized pigeonhole principle and symmetry, so are two of the people to the first, second, fourth and eighth to the right. Now there are lots of possibilities for that fourth person. You now have the hard work of using symmetry and the definition of who is acquainted with whom to eliminate all possible combinations of four people. Then you have to think about nonacquaintances.

86.a. What is the definition of $R(n,n)$?

86.b. If you average a bunch of numbers and each one is bigger than one, what can you say about the average?

86.c. Note that there are $2^{\binom{n}{2}}$ graphs on a set of n vertices.

86.d. A notation for the sum over all colorings c of K_m is

$$\sum_{c:c \text{ is a coloring of } K_m},$$

and a notation for the sum over all subsets N of M that have size n is

$$\sum_{N:N\subseteq M,\, |N|=n}$$

86.e. If you interchange the order of summation so that you sum over subsets first and colorings second, you can take advantage of the fact that for a fixed subset N, you can count count the number of colorings in which it is monochromatic.

86.f. You have an inequality involving m and n that tells you that $R(n,n) > m$. Suppose you could work with that inequality in order to show that if the inequality holds, then m is bigger than something. What could you conclude about $R(n,n)$?

87. Remember, a subset of $[n]$ either does or doesn't contain n.

90.b. A first order recurrence for a_n gives us a_n as a function of a_{n-1}.

91. Suppose you already knew the number of moves needed to solve the puzzle with $n-1$ rings.

92. If we have $n-1$ circles drawn in such a way that they define r_{n-1} regions, and we draw a new circle, each time it crosses another circle, except for the last time, it finishes dividing one region into two parts and starts dividing a new region into two parts.

Additional Hint: Compare r_n with the number of subsets of an n-element set.

98. You might try working out the cases $n = 2, 3, 4$ and then look for a pattern. Alternately, you could write $a_{n-1} = ba_{n-2} + d$, substitute the right hand side of this expression into $a_n = ba_{n-1} + d$ to get a recurrence involving only a_{n-2}, and then repeat a similar process with a_{n-2} and perhaps a_{n-3} and see a pattern that is developing.

102.a. There are several ways to see how to do this problem. One is to draw pictures of graphs with one edge, two edges, three edges, perhaps four edges and figure out the sum of the degrees. Another is to ask what deleting an edge does to the sum of the degrees. Another is to ask what a given edge "contributes" to the sum of the degrees.

102.b. To make your inductive step, think about what happens to a graph if you delete an edge.

102.d. Suppose that instead of summing the degree of v over all vertices v, you sum some quantity defined for each edge e over all the edges.

103. Whatever you say should be consistent with what you already know about degrees of vertices.

108. What happens if you choose an edge and delete it, but not its endpoints?

109. One approach to the problem is to use facts that we already know about degrees, vertices and edges. Another approach is to try deleting an edge from a tree with more than one vertex and analyze the possible numbers of vertices of degree one in what is left over.

111. When you get to four and especially five vertices, draw all the unlabeled trees you can think of, and then figure out in how many different ways you can put labels on the vertices.

112.b. Do some examples.

112.c. Is it possible for a_1 to be equal to one of the b_js?

112.d. You have seen that the sequence b determines a_1. Does it determine any other a_js? If you knew all the a_js and all the b_js, could you reconstruct the tree? What are the possible values of b_1? b_j?

113. What vertex or vertices in the sequence $b_1, b_2, \ldots, b_{n-1}$ can have degree 1?

115. If a vertex has degree 1, how many times does it appear in the Prüfer code of the tree? What about a vertex of degree 2?

116. How many vertices appear exactly once in the Prüfer code of the tree and how many appear exactly twice?

118. Think of selecting one edge of the tree at a time. Given that you have chosen some edges and have a graph whose connected components are trees, what is a good way to choose the next edge? To prove your method correct, use contradiction by assuming there is a spanning tree tree with lower total cost.

Additional Hint: Think of selecting one edge of the tree at a time. But now do it in such a way that one connected component is a tree and the other connected components have just one vertex. What is a good way to make the component that is a tree into a tree with one more vertex? To prove your method works, use contradiction by assuming there is a spanning tree with lower total cost.

119.a. If you have a spanning tree of G that contains e, is the graph that results from that tree by contracting e still a tree?

122.c. If you decide to put it on a shelf that already has a book, you have two choices of where to put it on that shelf.

122.e. Among all the places you could put books, on all the shelves, how many are to the immediate left of some book? How many other places are there?

123. How can you make sure that each shelf gets at least one book before you start the process described in Problem 122?

124. What is the relationship between the number of ways to distribute identical books and the number of ways to distribute distinct books?

125. Look for a relationship between a multiset of shelves and a way of distributing identical books to shelves

126. Note that $\binom{n+k-1}{k} = \binom{n+k-1}{n-1}$. So we have to figure out how choosing either k elements or $n - 1$ elements out of $n + k - 1$ elements constitutes the choice of a multiset. We really have no idea what set of $n + k - 1$ objects to use, so why not use $[n + k - 1]$? If we choose $n - 1$ of these objects, there are k left over, the same number as the number of elements of our multiset. Since our multiset is supposed to be chosen from an n-element set, perhaps we should let the n-element set be $[n]$. From our choice of $n - 1$ numbers, we have to decide on the multiplicity of 1 through n. For example with $n = 4$ and $k = 6$, we have $n + k - 1 = 9$. Here, shown with underlines, is a selection of $3 = n - 1$ elements from $[9]$: $1, 2, \underline{3}, 4\underline{5}, 6, 7, \underline{8}, 9$. How do the underlined elements give us a multiset of size 6 chosen from an $[4]$-element set? In this case, 1 has multiplicity 2, 2 has multiplicity 1, 3 has multiplicity 2, and 4 has multiplicity 1.

127. A solution to the equations assigns a nonnegative number to each of $1, 2, \ldots, m$ so that the nonnegative numbers add to r. Does such an assignment have a combinatorial meaning?

128. Can you think of some way of guaranteeing that each recipient gets m objects (assuming $k \geq mn$) right at the beginning of the process of passing the objects out?

129. We already know how to place k distinct books onto n distinct shelves so that each shelf gets at least one. Suppose we replace the distinct books with identical ones. If we permute the distinct books before replacement, does that affect the final outcome? There are other ways to solve this problem.

130. Do you see a relationship between compositions and something else we have counted already?

131. If we line up k identical books, how many adjacencies are there in between books?

133. Imagine taking a stack of k books, and breaking it up into stacks to put into the boxes in the same order they were originally stacked. If you are going to use n boxes, in how many places will you have to break the stack up into smaller stacks, and how many ways can you do this?

    Additional Hint: How many different bookcase arrangements correspond to the same way of stacking k books into n boxes so that each box has at least one book?

134. The number of partitions of $[k]$ into n parts in which k is not in a block relates to the number of partitions of $k - 1$ into some number of blocks in a way that involves n. With this in mind, review how you proved Pascal's (recurrence) equation.

137. What if the question asked about six sandwiches and two distinct bags? How does having identical bags change the answer?

138. What are the possible sizes of parts?

139. Suppose we make a list of the k items. We take the first k_1 elements to be the blocks of size 1. How many elements do we need to take to get k_2 blocks of size two? Which elements does it make sense to choose for this purpose?

141. To see how many broken permutations of a k element set into n parts do not have k is a part by itself, ask yourself how many broken permutations of $[7]$ result from adding 7 to the one of the two permutations in the broken permutation $\{14, 2356\}$.

142.b. Here it is helpful to think about what happens if you delete the entire block containing k rather than thinking about whether k is in a block by itself or not.

143. You can think of a function as assigning values to the blocks of its partition. If you permute the values assigned to the blocks, do you always change the function?

144. The Prüfer code of a labeled tree is a sequence of $n - 2$ entries that must be chose from the vertices that do not have degree 1. The sequence can be

though of as a function from the set $[n-2]$ to the set of vertices that do not have degree 1. What is special about this function?

145. When you add the number of functions mapping onto J over all possible subsets J of N, what is the set of functions whose size you are computing?

148. What if the j_i's don't add to k?

    Additional Hint: Think about listing the elements of the k-element set and labeling the first j_1 elements with label number 1.

149. The sum principle will help here.

150. How are the relevant j_i's in the multinomial coefficients you use here different from the j_i's in the previous problem.

151. Think about how binomial coefficients relate to expanding a power of a binomial and note that the binomial coefficient $\binom{n}{k}$ and the multinomial coefficient $\binom{n}{k,n-k}$ are the same.

152.a. We have related Stirling numbers to powers n^k. How are binomial coefficients related to falling factorial powers?

152.b. In the equation $\sum_{j=0}^{n} n^{\underline{j}} S(k,j) = n^k$, we might try substituting x for n. However we don't know what $\sum_{j=0}^{x}$ means when x is a variable. Is there anything other than n that makes a suitable upper limit for the sum? (Think about what you know about $S(k,j)$.

153. For the last question, you might try taking advantage of the fact that $x = x + 1 - 1$.

154. What does induction have to do with Equation (3.1)?

    Additional Hint: What could you assume inductively about $x^{\underline{k-1}}$ if you were trying to prove $x^{\underline{k}} = \sum_{n=0}^{k} s(k,n) x^n$?

156.a. There is a solution for this problem similar to the solution to Problem 154.

156.b. Is the recurrence you got familiar?

156.d. Show that $(-x)^{\underline{k}} = (-1)^k x^{\overline{k}}$ and $(-x)^{\overline{k}} = (-1)^k x^{\underline{k}}$.

    Additional Hint: The first hint lets you write an equation for $(-1)^k x^{\underline{k}}$ as a rising factorial of something else and then use what you know about expressing rising factorials in terms of falling factorials, after which you have to convert back to factorial powers of x.

162. How can you start with a partition of k and make it into a new partition of $k+1$ that is guaranteed to have a part of size one, even if the original partition didn't?

163. Draw a line through the top-left corner and bottom-right corner of the topleft box.

164. The largest part of a partition is the maximum number of boxes in a row of its Young diagram. What does the maximum number of boxes in a column tell us?

165. Draw all self conjugate partitions of integers less than or equal to 8. Draw all partitions of integers less than or equal to 8 into distinct odd parts (many of these will have just one part). Now try to see how to get from one set of drawings to the other in a consistent way.

166. Draw the partitions of six into even parts. Draw the partitions of six into parts used an even number of times. Look for a relationship between one set of diagrams and the other set of diagrams. If you have trouble, repeat the process using 8 or even 10 in place of 6.

167. Draw a partition of ten into four parts. Assume each square has area one. Then draw a rectangle of area 40 enclosing your diagram that touches the top of your diagram, the left side of your diagram and the bottom of your diagram. How does this rectangle give you a partition of 30 into four parts?

168.c. Consider two cases, $m' > m$ and $m' = m$.

168.d. Consider two cases, $n' > n$ and $n' = n$.

169. Suppose we take two repetitions of this complementation process. What rows and columns do we remove from the diagram?

    Additional Hint: To deal with an odd number of repetitions of the complementation process, think of it as an even number plus 1. Thus ask what kind of partition gives us the partition of one into one part after this complementation process.

170. How many compositions are there of k into n parts? What is the maximum number of compositions that could correspond to a given partition of k into n parts?

171.a. These two operations do rather different things to the number of parts, and you can describe exactly what only one of the operations does. Think about the Young diagram.

171.b. Think about the Young diagram. In only one of the two cases can you give an exact answer to the question.

171.c. Here the harder part requires that, after removal, you consider a range of possible numbers being partitioned and that you give an upper bound on the part size. However it lets you describe the number of parts exactly.

171.d. One of the two sets of partitions of smaller numbers from the previous part is more amenable to finding a recurrence than the other. The resulting recurrence does not have just two terms though.

171.h. If there is a sum equal to zero, there may very well be a partition of zero.

172. How does the number of compositions of k into n distinct parts compare to the number of compositions of k into n parts (not necessarily distinct)? What do compositions have to do with partitions?

173. While you could simply display partitions of 7 into three parts and partitions of 10 into three parts, we hope you won't. Perhaps you could write down the partitions of 4 into two parts and the partitions of 5 into two distinct parts and look for a natural bijection between them. So the hope is that you will discover a bijection from the set of partitions of 7 into three parts and the partitions of 10 into three distinct parts. It could help to draw the Young diagrams of partitions of 4 into two parts and the partitions of 5 into two distinct parts.

174. In the case $k = 4$ and $n = 2$, we have $m = 5$. In the case $k = 7$ and $n = 3$, we have $m = 10$.

175. What can you do to a Young diagram for a partition of k into n distinct parts to get a Young diagram of a partition of $k - n$ into some number of distinct parts?

176. For any partition of k into parts λ_1, λ_2, etc. we can get a partition of k into odd parts by factoring the highest power of two that we can from each λ_i, writing $\lambda_i = \gamma_i \cdot 2_i^k$. Why is γ_i odd? Now partition k into 2^{k_1} parts of size γ_1, 2^{k_2} parts of size γ_2, etc. and you have a partition of k into odd parts.

177. Suppose we have a partition of k into distinct parts. If the smallest part, say m, is smaller than the number of parts, we may add one to each of the m largest parts and delete the smallest part, and we have changed the parity of the number of parts, but we still have distinct parts. On the other hand, suppose the smallest part, again say m, is larger than or equal to the number of parts. Then we can subtract 1 from each part larger than m, and add a part equal to the number of parts larger than m. This changes the parity of the number of parts, but if the second smallest part is $m + 1$, the resulting partition does not have distinct parts. Thus this method does not work. Further, if it did always work, the case $k \neq \frac{3j^2+j}{2}$ would be covered also. However you can modify this method by comparing m not to the total number of parts, but to the number of rows at the top of the Young diagram that differ by exactly one from the row above. Even in this situation, there are certain slight additional assumptions you need to make, so this hint leaves you a lot of work to do. (It is reasonable to expect problems because of that exceptional case.) However, it should lead you in a useful direction.

183. Substitute something for A, P and B in your formula from Problem 181.

184. For example, to get the cost of the fruit selection APB you would want to get $x^{20}x^{25}x^{30} = x^{75}$.

186. Consider the example with $n = 2$. Then we have two variables, x_1 and x_2. Forgetting about x_2, what sum says we either take x_1 or we don't? Forgetting about x_1, what sum says we either take x_2 or we don't? Now what product says we either take x_1 or we don't *and* we either take x_2 or we don't?

188. For the last two questions, try multiplying out something simpler first, say $(a_0 + a_1 x + a_2 x^2)(b_0 + b_1 x + b_2 x^2)$. If this problem seems difficult the part that seems to cause students the most problems is converting the expression they get for a product like this into summation notation. If you are having this kind of problem, expand the product $(a_0 + a_1 x + a_2 x^2)(b_0 + b_1 x + b_2 x^2)$ and then figure out what the coefficient of x^2 is. Try to write that in summation notation.

189. Write down the formulas for the coefficients of x^0, x^1, x^2 and x^3 in

$$\left(\sum_{i=0}^{n} a_i x^i\right)\left(\sum_{j=0}^{m} b_j x^j\right).$$

190. How is this problem different from Problem 189? Is this an important difference from the point of view of the coefficient of x^k?

191. If this problem appears difficult, the most likely reason is because the definitions are all new and symbolic. Focus on what it means for $\sum_{k=0}^{\infty} c_k x^k$ to be the generating function for ordered pairs of total value k. In particular, how do we get an ordered pair with total value k? What do we need to know about the values of the components of the ordered pair?

192.b. You might try applying the product principle for generating functions to an appropriate power of the generating function you got in the first part of this problem.

   Additional Hint: In Problem 125 you found a formula for the number of k-element multisets chosen from an n-element set. Suppose you use this formula for a_k in $\sum_{k=0}^{\infty} a_k x^k$. What do you get the generating function for?

195. While you could use calculus techniques, there is a much simpler approach. Note that $1 + x = 1 - (-x)$.

Additional Hint: Can you see a way to use Problem 194?

197. Look for a power of a polynomial to get started.

    Additional Hint: The polynomial referred to in the first hint is a quotient of two polynomials. The power of the denominator can be written as a power series.

198. Intepret Problem 197 in terms of multisets.

199.e. When you factor out $x_1 x_2 \cdots x_n$ from the enumerator of trees, the result is a sum of terms of degree $n - 2$. (The degree of $x_1^{i_1} x_2^{i_2} \cdots x_n^{i_n}$ is $i_1 + i_2 + \cdots + i_n$.)

    Additional Hint: Write down the picture (using xs) of a tree on five vertices with two vertices of degree one, of one with three vertices of degree one, and with four vertices of degree 1. Factor $x_1 x_2 x_3 x_4 x_5$ out of the picture and look at what is left. How is it related to your vertices of degree one? Now remove the vertices of degree 1 from the tree and write down the picture of the tree that remains. What is special about the vertices of degree 1 of that tree. (You can just barely learn something from this with five vertex trees, so you might want to experiment a bit with six or seven vertex trees.)

200. This is a good place to apply the product principle for picture enumerators.

201.a. The product principle for generating functions helps you break the generating function into a product of ten simpler ones.

201.b. m was 10 in the previous part of this problem.

202. Think about conjugate partitions.

203.a. Don't be afraid of writing down a product of infinitely many power series.

203.b. From the fifth factor on, there is no way to choose a q^i that has i nonzero and less than five from the factor.

203.d. Describe to yourself how to get the coefficient of a given power of q.

204. If infinitely many of the polynomials had a nonzero coefficient for q, would the product make any sense?

205. $(1+q^2+q^4)(1+q^3+q^9)$ is the generating function for partitions of an integer into at most two twos and at most two threes.

206. $(1+q^2+q^4)(1+q^3+q^9)$ is the generating function for partitions of an integer into at most two twos and at most two threes. (This is intentionally the same hint as in the previous problem, but it has a different point in this problem.)

207. In the power series $\sum_{j=0}^{\infty} q^{2ij}$, the $2ij$ has a different interpretation if you think of it as $(2i) \cdot j$ or if you think of it as $i \cdot (2j)$.

208. Note that
$$\frac{1-q^2}{1-q} \cdot \frac{1-q^4}{1-q^2} \cdot \frac{1-q^6}{1-q^3} \cdot \frac{1-q^8}{1-q^4} = \frac{(1-q^6)(1-q^8)}{(1-q)(1-q^3)}.$$

209. Note that $q^i + q^{3i} + q^{5i} + \cdots = q^i(1 + q^2 + q^4 + \cdots)$.

210.a. We want to calculate the number of partitions whose Young diagrams fit into a two by two square. These partitions have at most two parts and the parts have size at most two. Thus they are partitions of 1, 2, 3, or 4. However not all partitions of 3 or 4 have diagrams that fit into a two by two square. Try writing down the relevant diagrams.

210.b. They are the generating function for the number of partitions whose Young diagram fits into a rectangle $n-1$ units wide and 1 unit deep or into a rectangle 1 unit wide and $n-1$ units deep respectively.

210.c. How can you get a diagram of a partition counted by partition is counted by $\begin{bmatrix} m+n \\ n \end{bmatrix}_q$ from one whose partition is counted by $\begin{bmatrix} m+n \\ m \end{bmatrix}_q$?

210.e.iii. Think about geometric operations on Young Diagrams

210.f. How would you use the Pascal recurrence to prove the corresponding result for binomial coefficients?

210.g. For finding a bijection, think about lattice paths.

210.h. If you could prove $\begin{bmatrix} m+n \\ n \end{bmatrix}_q$ is a polynomial function of q, what would that tell you about how to compute the limit as q approaches -1?

Additional Hint: Try computing a table of values of $\begin{bmatrix} m+n \\ n \end{bmatrix}_q$ with $q = -1$ by using the recurrence relation. Make a pretty big table so you can see what is happening.

211.c. You may run into a product of the form $\sum_{i=0}^{\infty} a^i x^i \sum_{j=0}^{\infty} b^j x^j$. Note that in the product, the coefficient of x^k is $\sum_{i=0}^{k} a^i b^{k-i} = \sum_{i=0}^{k} \frac{a^i}{b^i}$.

214. Our recurrence becomes $a_n = a_{n-1} + a_{n-2}$.

217.
$$\frac{5x+1}{(x-3)(x-5)} = \frac{cx + 5c + dx - 3d}{(x-3)(x-5)}$$
gives us
$$5x = cx + dx$$

$$1 = 5c - 3d.$$

218. To have
$$\frac{ax+b}{(x-r_1)(x-r_2)} = \frac{c}{x-r_1} + \frac{d}{x-r_2}$$
we must have
$$cx - r_2c + dx - r_1d = ax + b.$$

221. You can save yourself a tremendous amount of frustrating algebra if you arbitrarily choose one of the solutions and call it r_1 and call the other solution r_2 and solve the problem using these algebraic symbols in place of the actual roots.[1] Not only will you save yourself some work, but you will get a formula you could use in other problems. When you are done, substitute in the actual values of the solutions and simplify.

222.a. Once again it will save a lot of tedious algebra if you use the symbols r_1 and r_2 for the solutions as in Problem 221 and substitute the actual values of the solutions once you have a formula for a_n in terms of r_1 and r_2.

222.d. Think about how the binomial theorem might help you.

224.a. A Catalan path could touch the x-axis several times before it reaches $(2n, 0)$. Its first touch can be any point $(2i, 0)$ between $(2, 0)$ and $(2n, 0)$. For the path to touch first at $(2i, 0)$, the path must start with an upstep and then proceed as a Dyck path from $(1, 1)$ to $(2i-1, 1)$. From there it must take a downstep. Can you see a bijection between such Dyck paths and Catalan paths of a certain kind?

224.b. Does the right-hand side of the recurrence remind you of some products you have worked with?

224.c.
$$\frac{1 \cdot 3 \cdot 5 \cdots (2i-3)}{i!} = \frac{(2i-2)!}{(i-1)!2^i i!}.$$

226. Try drawing a Venn Diagram.

228. Try drawing a Venn Diagram.

231.b. For each student, how big is the set of backpack distributions in which that student gets the correct backpack? It might be a good idea to first consider cases with $n = 3, 4$, and 5.

Additional Hint: For each pair of students (say Mary and Jim, for example) how big is the set of backpack distributions in which the students in this pair get the correct backpack. What does the question have to do with unions or

---
[1] We use the words roots and solutions interchangeably.

intersections of sets. Keep on increasing the number of students for which you ask this kind of question.

232. Try induction.

   Additional Hint: We can apply the formula of Problem 226 to get

   $$\left|\bigcup_{i=1}^{n} A_i\right| = \left|\left(\bigcup_{i=1}^{n-1} A_i\right) \cup A_n\right|$$
   $$= \left|\bigcup_{i=1}^{n-1} A_i\right| + |A_n| - \left|\left(\bigcup_{i=1}^{n-1} A_i\right) \cap A_n\right|$$
   $$= \left|\bigcup_{i=1}^{n-1} A_i\right| + |A_n| - \left|\bigcup_{i=1}^{n-1} A_i \cap A_n\right|$$

233.b. Let T be the set of all i such that $x \in A_i$. In terms of x, what is different about the i in T and those not in T?

   Additional Hint: You may come to a point where the binomial theorem would be helpful.

235. Notice that it is straightforward to figure out how many ways we may pass out the apples so that child i gets five or more apples: give five apples to child i and then pass out the remaining apples however you choose. And if we want to figure out how many ways we may pass out the apples so that a given set C of children each get five or more apples, we give five to each child in C and then pass out the remaining $k - 5|C|$ apples however we choose.

236. Start with two questions that can apply to any inclusion-exclusion problem. Do you think you would be better off trying to compute the size of a union of sets or the size of a complement of a union of sets? What kinds of sets (that are conceivably of use to you) is it easy to compute the size of? (The second question can be interpreted in different ways, and for each way of interpreting it, the answer may help you see something you can use in solving the problem.)

   Additional Hint: Suppose we have a set S of couples whom we want to seat side by side. We can think of lining up $|S|$ couples and $2n - 2|S|$ individual people in a circle. In how many ways can we arrange this many items in a circle?

237. Reason somewhat as you did in Problem 236, noting that if the set of couples who do sit side-by-side is nonempty, then the sex of the person at each place at the table is determined once we seat one couple in that set.

   Additional Hint: Think in terms of the sets A_i of arrangements of people in which couple i sits side-by-side. What does the union of the sets A_i have to do with the problem?

239. What does Problem 238 have to do with this question?

242.a. For each edge in F to connect two vertices of the same color, we must have all the vertices in a connected component of the graph with vertex set V and edge set F colored the same color.

242.c. How does the number you are trying to compute relate to the union of the sets A_i?

243. One way to get a proper coloring of $G - e$ is to start with a proper coloring of G and remove e. But there are other colorings of G that become proper when you remove e.

246. One approach would be to try to guess the result by doing a bunch of examples and use induction to prove you are right. If you try this, what will you be able to use to make the induction step work? There are other approaches as well.

253.a. What do you want $\varphi^n \circ \varphi^{-1}$ to be?

254. If $\sigma^i = \sigma^j$ and $i \neq j$, what can you conclude about ι?

256.b. What does it mean for one function to be the inverse of another one?

261. Once you know where the corners of the square go under the action of an isometry, how much do you know about the isometry?

264. In how many ways can you choose a place to which you can move vertex 1? Having done that, in how many ways can you place the three vertices adjacent to vertex 1?

265.a. In how many ways can you choose a place to which you can move vertex 1? Having done that, in how many ways can you place the three vertices adjacent to vertex 1?

265.b. Why is it sufficient to focus on permutations that take vertex 1 to itself?

270. If a subgroup contains, say, ρ^3 and some flip, how many elements of D_4 must it contain?

272. If the list $(i \ \sigma(i) \ \sigma^2(i) \ \ldots \ \sigma^n(i))$ does not have repeated elements but the list $(i \ \sigma(i) \ \sigma^2(i) \ \ldots \ \sigma^n(i) \ \sigma^{n+1}(i))$ does have repeated elements, then which element or elements are repeats?

277. The element k is either in a cycle by itself or it isn't.

286. Before you try to show that $\bar{\bar{\sigma}}$ actually is a permutation of the colorings, it would be useful to verify the second part of the definition of a group action, namely that $\bar{\bar{\sigma}} \circ \bar{\bar{\varphi}} = \overline{\overline{\sigma \circ \varphi}}$.

289. If $z \in Gx$ and $z \in Gy$, how can you use elements of G to explain the relationship between x and y?

    Additional Hint: Suppose σ is a fixed member of G. As τ ranges over G, which elements of G occur as $\tau\sigma$?

295. How does the size of a multiorbit compare to the size of G?

301. We are asking for the number of orbits of some group on lists of four Rs, six Bs, and seven Gs.

305. There are five kinds of elements in the rotation group of the cube. For example, there are six rotations by 90 degrees or 270 degrees around an axis connecting the centers of two opposite faces and there are 8 rotations (of 120 degrees and 240 degrees, respectively) around an axis connecting two diagonally opposite vertices.

306. Is it possible for a nontrivial rotation to fix any coloring?

309. There are 48 elements in the group of automorphisms of the graph.

    Additional Hint: For this problem, it may be easier to ask which group elements fix a coloring rather than which colorings are fixed by a group element.

326. The group of automorphisms of the graph has 48 elements and contains D_6 as a subgraph.

    Additional Hint: The permutations with four one-cycles and the two-cycle (1 4), (2 5), or (3 6) are in the group of automorphisms. Once you know the cycle structure of D_6 and $(1\ 4)D_6 = \{(1\ 4)\sigma | \sigma \in D_6\}$, you know the cycle structure of every element of the group.

327. What does the symmetric group on five vertices have to do with this problem?

329.c. In the relation of a function, how many pairs $(x, f(x))$ have the same x-value?

332. For the second question, how many arrows have to leave the empty set? How many arrows have to leave a set of size one?

339. What is the domain of $g \circ f$?

345. If we have scoops of vanilla, chocolate, and strawberry sitting in a circle in a dish, can we distinguish between VCS and VSC?

349. To show a relation is an equivalence relation, you need to show it satisfies the definition of an equivalence relation.

353. To get you started, in Problem 38 the equivalence classes correspond to seating arrangements.

361. You've probably guessed that the sum is n^2. To prove this by contradiction, you have to assume it is false, that is, that there is an n such that $1 + 3 + 5 + \cdots + 2n - 1 \neq n^2$. Then the method of Problem 360 says there must be a smallest such n and suggests we call it k. Why do you know that $1 + 3 + 5 + \cdots + 2k - 3 = (k-1)2$? What happens if you add $2n - 1$ to both sides of the equation?

365. You've probably already seen that, with small values of n, sometimes n^2 and sometimes 2^n is bigger. But if you keep experimenting one of the functions seems to get bigger and stay bigger than the other. The number $n = b$ where this change occurs is a good choice for a base case. So as not to spoil the problem for you, we won't say here what this value of b is. However you shouldn't be surprised later in the proof if you need to use the assumption that n/gtb.

Additional Hint: You may have reached the point of assuming that $2^{k-1} > (k-1)^2$ and found yourself wondering how to prove that $2^k > k^2$. A natural thing to try is multiplying both sides of $2^{k-1} > (k-1)^2$ by 2. This ends up giving you $2^k > 2k^2 - 4k + 2$. Based on previous experience it is natural for you to expect to see how to turn this new right hand side into k^2 but not see how to do it. Here is the hint. You only need to show that the right hand side is greater than or equal to k^2. For this purpose you need to show that one of the two k^2s in $2k^2$ somehow balances out the $-4k$. See if you can figure out how the fact that you are only considering ks with $k > b$ can help you out.

366. When you suspect an argument is not valid, it may be helpful to explicitly try several values of n to see if it makes sense for them. Often small values of n are adequate to find the flaw. If you find one flaw, it invalidates everything that comes afterwards (unless, of course, you can fix the flaw).

370. You might start out by ignoring the word unique and give a proof of the simpler theorem that results. Then look at your proof to see how you can include uniqueness in it.

377. An earlier problem may help you put your answer into a simpler form.

378. What is the power series representation of e^{x^2}?

381. There is only one element that you may choose. In the first case you either choose it or you don't.

387.b. At some point, you may find the binomial theorem to be useful.

391. Notice that any permutation is a product of a derangement of the elements not fixed by the permutation times a permutation whose cycle decomposition consists of one-cycles.

392. A binomial coefficient is likely to appear in your answer.

397. If $f(x) = \sum_{i=0}^{\infty} a_i \frac{x^i}{i!}$ and $g(x) = \sum_{j=0}^{\infty} b_j \frac{x^j}{j!}$, what is the coefficient of $\frac{x^n}{n!}$ in $f(x)g(x)$? don't be surprised if your answer has a binomial coefficient in it. In fact, the binomial coefficient should help you finish the problem.

399. Since the sets of a k-set structure are nonempty and disjoint, the k-element set of sets can be arranged as a k-tuple in $k!$ ways.

403. The alternate definition of a funciton in Section 3.1.2 can be restated to say that a function from a k-element set K to an n-element set N can be thought of as an n-tuple of sets, perhaps with some empty, whose union is K. In order to think of the function as an n-tuple, we number the elements of N as number 1 through number n. Then the ith set in the n-tuple is the set of elements mapped to the ith element of N in our numbering?

404. Don't be surprised if you see a hyperbolic sine or hyperbolic cosine in your answer. If you aren't familiar with these functions, look them up in a calculus book.

407. The EGF for $\sum_{i=1}^{n} \binom{n}{k} k$ is $\sum_{n=1}^{\infty} \sum_{i=1}^{n} \frac{n!}{k!(n-k)!} k \frac{x^n}{n!}$. You can cancel out the $n!$ terms and the k terms. Now try to see if what is left can be regarded as the product of two EGFs.

421.a. To apply the exponential formula, we must take the exponential function of an EGF whose constant term is zero, or in other words, for a species of structures that has no structures that use the empty set.

421.b. Once you know the vertex set of a graph, all you have to do to specify the graph is to specify its set of edges.

421.d. What is the calculus definition of $\log(1+y)$?

421.f. Look for a formula that involves summing over all partitions of the integer n.

Appendix E

GNU Free Documentation License

Version 1.3, 3 November 2008
 Copyright © 2000, 2001, 2002, 2007, 2008 Free Software Foundation, Inc. `<http://www.fsf.org/>`
 Everyone is permitted to copy and distribute verbatim copies of this license document, but changing it is not allowed.

0. PREAMBLE The purpose of this License is to make a manual, textbook, or other functional and useful document "free" in the sense of freedom: to assure everyone the effective freedom to copy and redistribute it, with or without modifying it, either commercially or noncommercially. Secondarily, this License preserves for the author and publisher a way to get credit for their work, while not being considered responsible for modifications made by others.

This License is a kind of "copyleft", which means that derivative works of the document must themselves be free in the same sense. It complements the GNU General Public License, which is a copyleft license designed for free software.

We have designed this License in order to use it for manuals for free software, because free software needs free documentation: a free program should come with manuals providing the same freedoms that the software does. But this License is not limited to software manuals; it can be used for any textual work, regardless of subject matter or whether it is published as a printed book. We recommend this License principally for works whose purpose is instruction or reference.

1. APPLICABILITY AND DEFINITIONS This License applies to any manual or other work, in any medium, that contains a notice placed by the copyright holder saying it can be distributed under the terms of this License. Such a notice grants a world-wide, royalty-free license, unlimited in duration, to use that work under the conditions stated herein. The "Document", below, refers to any such manual or work. Any member of the public is a licensee, and is addressed as "you". You accept the license if you copy, modify or distribute the work in a way requiring permission under copyright law.

A "Modified Version" of the Document means any work containing the Document or a portion of it, either copied verbatim, or with modifications and/or translated into another language.

A "Secondary Section" is a named appendix or a front-matter section of the Document that deals exclusively with the relationship of the publishers or authors of the Document to the Document's overall subject (or to related matters) and contains nothing that could fall directly within that overall subject. (Thus, if the Document is in part a textbook of mathematics, a Secondary Section may not explain any mathematics.) The relationship could be a matter of historical connection with the subject or with related matters, or of legal, commercial, philosophical, ethical or political position regarding them.

The "Invariant Sections" are certain Secondary Sections whose titles are designated, as being those of Invariant Sections, in the notice that says that the Document is released under this License. If a section does not fit the above definition of Secondary then it is not allowed to be designated as Invariant.

The Document may contain zero Invariant Sections. If the Document does not identify any Invariant Sections then there are none.

The "Cover Texts" are certain short passages of text that are listed, as Front-Cover Texts or Back-Cover Texts, in the notice that says that the Document is released under this License. A Front-Cover Text may be at most 5 words, and a Back-Cover Text may be at most 25 words.

A "Transparent" copy of the Document means a machine-readable copy, represented in a format whose specification is available to the general public, that is suitable for revising the document straightforwardly with generic text editors or (for images composed of pixels) generic paint programs or (for drawings) some widely available drawing editor, and that is suitable for input to text formatters or for automatic translation to a variety of formats suitable for input to text formatters. A copy made in an otherwise Transparent file format whose markup, or absence of markup, has been arranged to thwart or discourage subsequent modification by readers is not Transparent. An image format is not Transparent if used for any substantial amount of text. A copy that is not "Transparent" is called "Opaque".

Examples of suitable formats for Transparent copies include plain ASCII without markup, Texinfo input format, LaTeX input format, SGML or XML using a publicly available DTD, and standard-conforming simple HTML, PostScript or PDF designed for human modification. Examples of transparent image formats include PNG, XCF and JPG. Opaque formats include proprietary formats that can be read and edited only by proprietary word processors, SGML or XML for which the DTD and/or processing tools are not generally available, and the machine-generated HTML, PostScript or PDF produced by some word processors for output purposes only.

The "Title Page" means, for a printed book, the title page itself, plus such following pages as are needed to hold, legibly, the material this License requires to appear in the title page. For works in formats which do not have any title page as such, "Title Page" means the text near the most prominent appearance of the work's title, preceding the beginning of the body of the text.

The "publisher" means any person or entity that distributes copies of the Document to the public.

A section "Entitled XYZ" means a named subunit of the Document whose title either is precisely XYZ or contains XYZ in parentheses following text that translates XYZ in another language. (Here XYZ stands for a specific section name mentioned below, such as "Acknowledgements", "Dedications", "Endorsements", or "History".) To "Preserve the Title" of such a section when you modify the Document means that it remains a section "Entitled XYZ" according to this definition.

The Document may include Warranty Disclaimers next to the notice which states that this License applies to the Document. These Warranty Disclaimers are considered to be included by reference in this License, but only as regards disclaiming warranties: any other implication that these Warranty Disclaimers may have is void and has no effect on the meaning of this License.

2. VERBATIM COPYING You may copy and distribute the Document in any medium, either commercially or noncommercially, provided that this License, the copyright notices, and the license notice saying this License applies to the Document are reproduced in all copies, and that you add no other conditions whatsoever to those of this License. You may not use technical measures to obstruct or control the reading or further copying of the copies you make or distribute. However, you may accept compensation in exchange for copies. If you distribute a large enough number of copies you must also follow the conditions in section 3.

You may also lend copies, under the same conditions stated above, and you may publicly display copies.

3. COPYING IN QUANTITY If you publish printed copies (or copies in media that commonly have printed covers) of the Document, numbering more than 100, and the Document's license notice requires Cover Texts, you must enclose the copies in covers that carry, clearly and legibly, all these Cover Texts: Front-Cover Texts on the front cover, and Back-Cover Texts on the back cover. Both covers must also clearly and legibly identify you as the publisher of these copies. The front cover must present the full title with all words of the title equally prominent and visible. You may add other material on the covers in addition. Copying with changes limited to the covers, as long as they preserve the title of the Document and satisfy these conditions, can be treated as verbatim copying in other respects.

If the required texts for either cover are too voluminous to fit legibly, you should put the first ones listed (as many as fit reasonably) on the actual cover, and continue the rest onto adjacent pages.

If you publish or distribute Opaque copies of the Document numbering more than 100, you must either include a machine-readable Transparent copy along with each Opaque copy, or state in or with each Opaque copy a computer-network location from which the general network-using public has access to download using public-standard network protocols a complete Transparent copy of the Document, free of added material. If you use the latter option, you must take reasonably prudent steps, when you begin distribution of Opaque copies in quantity, to ensure that this Transparent copy will remain thus accessible at the stated location until at least one year after the last time you distribute an Opaque copy (directly or through your agents or retailers) of that edition to the public.

It is requested, but not required, that you contact the authors of the Document well before redistributing any large number of copies, to give them a chance to provide you with an updated version of the Document.

4. MODIFICATIONS You may copy and distribute a Modified Version of the Document under the conditions of sections 2 and 3 above, provided that you release the Modified Version under precisely this License, with the Modified Version filling the role of the Document, thus licensing distribution and modification of the Modified Version to whoever possesses a copy of it. In addition, you must do these things in the Modified Version:

- A. Use in the Title Page (and on the covers, if any) a title distinct from that of the Document, and from those of previous versions (which should, if there were any, be listed in the History section of the Document). You may use the same title as a previous version if the original publisher of that version gives permission.
- B. List on the Title Page, as authors, one or more persons or entities responsible for authorship of the modifications in the Modified Version, together with at least five of the principal authors of the Document (all of its principal authors, if it has fewer than five), unless they release you from this requirement.
- C. State on the Title page the name of the publisher of the Modified Version, as the publisher.
- D. Preserve all the copyright notices of the Document.
- E. Add an appropriate copyright notice for your modifications adjacent to the other copyright notices.
- F. Include, immediately after the copyright notices, a license notice giving the public permission to use the Modified Version under the terms of this License, in the form shown in the Addendum below.
- G. Preserve in that license notice the full lists of Invariant Sections and required Cover Texts given in the Document's license notice.
- H. Include an unaltered copy of this License.
- I. Preserve the section Entitled "History", Preserve its Title, and add to it an item stating at least the title, year, new authors, and publisher of the Modified Version as given on the Title Page. If there is no section Entitled "History" in the Document, create one stating the title, year, authors, and publisher of the Document as given on its Title Page, then add an item describing the Modified Version as stated in the previous sentence.
- J. Preserve the network location, if any, given in the Document for public access to a Transparent copy of the Document, and likewise the network locations given in the Document for previous versions it was based on. These may be placed in the "History" section. You may omit a network location for a work that was published at least four years before the Document itself, or if the original publisher of the version it refers to gives permission.
- K. For any section Entitled "Acknowledgements" or "Dedications", Preserve the Title of the section, and preserve in the section all the substance and tone of each of the contributor acknowledgements and/or dedications given therein.
- L. Preserve all the Invariant Sections of the Document, unaltered in their text and in their titles. Section numbers or the equivalent are not considered part of the section titles.
- M. Delete any section Entitled "Endorsements". Such a section may not be included in the Modified Version.
- N. Do not retitle any existing section to be Entitled "Endorsements" or to conflict in title with any Invariant Section.
- O. Preserve any Warranty Disclaimers.

If the Modified Version includes new front-matter sections or appendices that qualify as Secondary Sections and contain no material copied from the Document, you may at your option designate some or all of these sections as invariant. To do this, add their titles to the list of Invariant Sections in the Modified Version's license notice. These titles must be distinct from any other section titles.

You may add a section Entitled "Endorsements", provided it contains nothing but endorsements of your Modified Version by various parties — for example, statements of peer review or that the text has been approved by an organization as the authoritative definition of a standard.

You may add a passage of up to five words as a Front-Cover Text, and a passage of up to 25 words as a Back-Cover Text, to the end of the list of Cover Texts in the Modified Version. Only one passage of Front-Cover Text and one of Back-Cover Text may be added by (or through arrangements made by) any one entity. If the Document already includes a cover text for the same cover, previously added by you or by arrangement made by the same entity you are acting on behalf of, you may not add another; but you may replace the old one, on explicit permission from the previous publisher that added the old one.

The author(s) and publisher(s) of the Document do not by this License give permission to use their names for publicity for or to assert or imply endorsement of any Modified Version.

5. COMBINING DOCUMENTS You may combine the Document with other documents released under this License, under the terms defined in section 4 above for modified versions, provided that you include in the combination all of the Invariant Sections of all of the original documents, unmodified, and list them all as Invariant Sections of your combined work in its license notice, and that you preserve all their Warranty Disclaimers.

The combined work need only contain one copy of this License, and multiple identical Invariant Sections may be replaced with a single copy. If there are multiple Invariant Sections with the same name but different contents, make the title of each such section unique by adding at the end of it, in parentheses, the name of the original author or publisher of that section if known, or else a unique number. Make the same adjustment to the section titles in the list of Invariant Sections in the license notice of the combined work.

In the combination, you must combine any sections Entitled "History" in the various original documents, forming one section Entitled "History"; likewise combine any sections Entitled "Acknowledgements", and any sections Entitled "Dedications". You must delete all sections Entitled "Endorsements".

6. COLLECTIONS OF DOCUMENTS You may make a collection consisting of the Document and other documents released under this License, and replace the individual copies of this License in the various documents with a single copy that is included in the collection, provided that you follow the rules of this License for verbatim copying of each of the documents in all other respects.

You may extract a single document from such a collection, and distribute it individually under this License, provided you insert a copy of this License into the extracted document, and follow this License in all other respects regarding verbatim copying of that document.

7. AGGREGATION WITH INDEPENDENT WORKS A compilation of the Document or its derivatives with other separate and independent documents or works, in or on a volume of a storage or distribution medium, is called an "aggregate" if the copyright resulting from the compilation is not used to limit the legal rights of the compilation's users beyond what the individual works permit. When the Document is included in an aggregate, this License does not apply to the other works in the aggregate which are not themselves derivative works of the Document.

If the Cover Text requirement of section 3 is applicable to these copies of the Document, then if the Document is less than one half of the entire aggregate, the Document's Cover Texts may be placed on covers that bracket the Document within the aggregate, or the electronic equivalent of covers if the Document is in electronic form. Otherwise they must appear on printed covers that bracket the whole aggregate.

8. TRANSLATION Translation is considered a kind of modification, so you may distribute translations of the Document under the terms of section 4. Replacing Invariant Sections with translations requires special permission from their copyright holders, but you may include translations of some or all Invariant Sections in addition to the original versions of these Invariant Sections. You may include a translation of this License, and all the license notices in the Document, and any Warranty Disclaimers, provided that you also include the original English version of this License and the original versions of those notices and disclaimers. In case of a disagreement between the translation and the original version of this License or a notice or disclaimer, the original version will prevail.

If a section in the Document is Entitled "Acknowledgements", "Dedications", or "History", the requirement (section 4) to Preserve its Title (section 1) will typically require changing the actual title.

9. TERMINATION You may not copy, modify, sublicense, or distribute the Document except as expressly provided under this License. Any attempt otherwise to copy, modify, sublicense, or distribute it is void, and will automatically terminate your rights under this License.

However, if you cease all violation of this License, then your license from a particular copyright holder is reinstated (a) provisionally, unless and until the copyright holder explicitly and finally terminates your license, and (b) permanently, if the copyright holder fails to notify you of the violation by some reasonable means prior to 60 days after the cessation.

Moreover, your license from a particular copyright holder is reinstated permanently if the copyright holder notifies you of the violation by some reasonable means, this is the first time you have received notice of violation of this License (for any work) from that copyright holder, and you cure the violation prior to 30 days after your receipt of the notice.

Termination of your rights under this section does not terminate the licenses of parties who have received copies or rights from you under this License. If your rights have been terminated and not permanently reinstated, receipt of a copy of some or all of the same material does not give you any rights to use it.

10. FUTURE REVISIONS OF THIS LICENSE The Free Software Foundation may publish new, revised versions of the GNU Free Documentation License from time to time. Such new versions will be similar in spirit to the present version, but may differ in detail to address new problems or concerns. See `http://www.gnu.org/copyleft/`.

Each version of the License is given a distinguishing version number. If the Document specifies that a particular numbered version of this License "or any later version" applies to it, you have the option of following the terms and conditions either of that specified version or of any later version that has been published (not as a draft) by the Free Software Foundation. If the Document does not specify a version number of this License, you may choose any version ever published (not as a draft) by the Free Software Foundation. If the Document specifies that a proxy can decide which future versions of this License can be used, that proxy's public statement of acceptance of a version permanently authorizes you to choose that version for the Document.

11. RELICENSING "Massive Multiauthor Collaboration Site" (or "MMC Site") means any World Wide Web server that publishes copyrightable works and also provides prominent facilities for anybody to edit those works. A public wiki that anybody can edit is an example of such a server. A "Massive Multiauthor Collaboration" (or "MMC") contained in the site means any set of copyrightable works thus published on the MMC site.

"CC-BY-SA" means the Creative Commons Attribution-Share Alike 3.0 license published by Creative Commons Corporation, a not-for-profit corporation with a principal place of business in San Francisco, California, as well as future copyleft versions of that license published by that same organization.

"Incorporate" means to publish or republish a Document, in whole or in part, as part of another Document.

An MMC is "eligible for relicensing" if it is licensed under this License, and if all works that were first published under this License somewhere other than this MMC, and subsequently incorporated in whole or in part into the MMC, (1) had no cover texts or invariant sections, and (2) were thus incorporated prior to November 1, 2008.

The operator of an MMC Site may republish an MMC contained in the site under CC-BY-SA on the same site at any time before August 1, 2009, provided the MMC is eligible for relicensing.

Index

$S(k,n)$, 58
Π notation, 8
$n!$, 8
    Stirling's formula for, 19
$n^{\overline{k}}$, 55
$n^{\underline{k}}$, 8
q-ary factorial, 84
q-binomial coefficient, 84
$s(k,n)$, 62

action of a group on a set, 115
arithmetic progression, 39
arithmetic series, 40
associative law, 105
asymptotic combinatorics, 36
automorphism (of a graph), 124, 131

basis (for polynomials), 61
Bell Number, 59
bijection, 11, 137
bijection principle, 11
binomial coefficient, 11
    q-binomial, 84
Binomial Theorem, 24
binomial theorem
    extended, 80
block of a partition, 6, 141
broken permutation, 57
Burnside's Lemma, 123

Cartesian product, 5
Catalan Number, 22, 124
    recurrence for, 89, 90
Catalan number
    generating function for, 89
Catalan Path, 22

Cauchy-Frobenius-Burnside
    Theorem, 123
characteristic function, 14
chromatic polynomial of a graph, 99
Chung-Feller Theorem, 23
closure property, 105
coefficient
    multinomial, 60
coloring
    standard notation, 117
    standard ordering, 117
coloring of a graph, 98
    proper, 98
combinations, 11
commutative law, 111
complement, 96
complement of a partition, 66
composition, 28, 137
    k parts, 28
    number of, 28
composition of functions, 104
compositions
    number of, 28
congruence modulo n, 140
conjugate of an integer partition, 65
connected component graph, 165
connected component of a graph, 98, 165
connected structures and EGFs, 163
constant coefficient linear
    recurrence, 40
contraction, 47
cost of a spanning tree, 46
cycle (in a graph), 43
cycle (of a permutation), 112
    element of, 112

199

equivalent, 112
cycle index, 129
cycle monomial, 129
cyclic group, 113

definition
    inductive, 33
    recursive, 33
degree of a vertex, 42
degree sequence, 50, 72
    ordered, 50
deletion, 47
deletion-contraction recurrence, 48, 99
derangement, 95
derangement problem, 95
diagram
    of a partition
        Ferrers, 64
        Young, 64
digraph, 9, 135
dihedral group, 107
Dijkstra's algorithm, 48
directed graph, 9, 135
disjoint, 4
    multisets, 121
distance in a graph, 48
distance in a weighted graph, 48
domain (of a function), 133
double induction, 35
    strong, 35
driving function, 40
Dyck path, 22

edge, 26, 41, 135
    in a digraph, 135
    of a complete graph, 26
EGF, 152
enumerator
    fixed point, 127
    orbit, 126
equivalence class, 141
equivalence relation, 140, 142
equivalent cycle, 112
exponential formula, 163
    connected structures for, 165
exponential generating function, 152
exponential generating functions
    product principle for, 160
exponential generating functions for connected structures, 163
extended binomial theorem, 80

F-structures, 158
factorial, 8, 33, 55
    q-ary, 84
    falling, 55
factorial power
    falling, 8
    rising, 55
falling factorial power, 8, 55
Ferrers diagram, 64
Fibonacci numbers, 86, 88
fix, 121
fixed point enumerator, 127
function, 3, 133
    alternate definition, 55
    bijection, 11
    characteristic, 14
    composition, 137
    digraph of, 9
    driving, 40
    identity, 104
    injection, 4
    inverse, 105
    one-to-one, 4, 134
    onto, 10, 134
        and Stirling Numbers, 60
    ordered, 55
        onto, 55
    relation of, 133
    surjection, 10, 134
functions
    composition of, 104
    number of, 53
    one-to-one
        number of, 53
    onto
        number of, 97

general product principle, 6, 34
generating function, 76
    exponential, 152
        product principle for, 160
    ordinary, 152
    product principle for, 79

geometric progression, 40
geometric series, 40, 79
graph, 41
    chromatic polynomial of, 99
    coloring of, 98
        proper, 98
    complete, 26
    connected component of, 98, 165
    coordinate, 135
    directed, 9, 135
    distance in, 48
    simple, 165
graphs
    isomorphic, 131
Gray Code, 28
greedy method, 46
group acting on a set, 115
group action on colorings, 118
group of permutations, 105

hatcheck problem, 95
homogeneous linear recurrence, 40

identity function, 104, 137
identity property, 105
identity property (for permutations), 105
inclusion and exclusion principle, 93
    for unions of sets, 95
indicator polynomials, 151
induction
    double, 35
    mathematical, the principle of, 31, 147
    mathematical, the strong principle of, 32
    strong double, 35
inductive
    conclusion, 32
    hypothesis, 32
    step, 32
inductive definition, 33
injection, 4, 134
inverse function, 105
inverse property, 105
involution, 114
isometry, 108
isomorphic

graphs, 131

k-set structures, 160

Lah number, 57
lattice path, 20
    diagonal, 20
length (of a path), 48
linear recurrence, 40, 87
    constant coefficient, 40
    homogeneous, 40
    second order, 86

mathematical induction
    double, 35
    principle of, 31, 147
    strong double, 35
ménage problem, 97
method
    probabilistic, 36
minimum cost spanning tree, 46
monochromatic subgraph, 37
multinomial coefficient, 60
multiorbit, 120
multiorbits, 127
multiplicity in a multiset, 56
multiset, 56
multisets
    product principle, 122
    quotient principle, 122
    sum principle, 121
    union, 121

one-to-one, 4
one-to-one function, 134
onto function, 10, 134
    counting, 60
    ordered, 55
onto functions
    number of, 97
orbit, 119
orbit enumerator, 126
Orbit-Fixed Point Theorem, 127
ordered degree sequence, 72
ordered function, 55
ordered onto function, 55
ordered pair, 3
ordinary generating function, 152

pair structure, 159
pair,ordered, 3
partial fractions
    method of, 88
partition
    blocks of, 6
    of a set, 5, 58
        Stirling Numbers, 58
partition (of a set), 141
partition of a set
    type vector, 59
partition of an integer, 63
    conjugate of, 65
    decreasing list, 64
    Ferrers diagram, 64
    into n parts, 63
    self conjugate, 65
    type vector, 64
    Young diagram, 64
partitions of a set
    number of, 59
Pascal's Triangle, 12
path
    in graph, 43
    lattice, 20
        diagonal, 20
    length of, 48
permutation
    k-element, 8
    as a bijection, 11
    broken, 57
    cycle of, 112
    two row notation, 107
permutation group, 105
picture enumerator, 74
picture enumerators
    product principle for, 75
pigeonhole principle, 25
    generalized, 26
Pólya-Redfield Theorem, 129
principle
    bijection, 11
    product, 5, 6
        general, 6
    quotient, 143
    sum, 5, 6

principle of inclusion and exclusion, 93
    for unions of sets, 95
principle of mathematical induction, 31, 147
probabilistic method, 36
product
    Cartesian, 5
product notation, 8
product principle, 5, 6
    for multisets, 122
    general, 6, 34
    picture enumerators, 75
product principle for exponential generating functions, 160
product principle for generating functions, 79
progression
    arithmetic, 39
    geometric, 40
proper coloring of a graph, 98

quotient principle, 18, 143
    for multisets, 122

range (of a function), 133
recurrence, 38
    constant coefficient, 86, 87
    deletion-contraction, 48
    linear, 40, 86, 87
    linear homogeneous, 40
    second order, 86, 87
    solution to, 38
    two variable, 58
recurrence relation, 38
recursive definition, 33
reflexive, 139
relation, 133
    equivalence, 140, 142
    of a function, 133
    recurrence, 38
    reflexive, 139, 140
    transitive, 140
rising factorial power, 55
rotation group, 105

second order recurrence, 86
self-conjugate partition, 65

sequence
    degree, 50
series
    arithmetic, 40
    geometric, 40, 79
set
    colorings of action of a group
        on, 116
sets
    disjoint, 4
    mutually disjoint, 4
simple graph, 165
space of polynomials, 61
spanning tree, 45
    cost of, 46
    minimum cost, 46
species, 158
    exponential generating function
        for, 159
standard notation for a coloring, 117
Stirling Number
    first kind, 62
    second kind, 58, 97
Stirling's formula for $n!$, 19
Stirling's triangle
    first kind, 62
    second kind, 58
strong double induction, 35
strong principle of mathematical
    induction, 32
structure, 158
    pair, 159

using a set, 158
subgroup, 111
sum principle, 5, 6, 93
    for multisets, 121
surjection, 10, 134
surjections
    number of, 97
symmetric, 140
symmetric group, 106

transitive, 140
tree, 43
    spanning, 45
        cost of, 46
        minimum cost, 46
Twentyfold Way, 52
two row notation, 107
type vector for a partition of an
    integer, 64
type vector of a partition of a set, 59

union of multisets, 121
uses
    a structure using a set, 158

vertex, 26, 41, 135
    degree of, 42
    of a complete graph, 26, 135

walk, 43

Young diagram, 64

Printed in Great Britain
by Amazon